BUILD YOUR OWN WORKING
FIBEROPTIC
INFRARED AND LASER
SPACE-AGE PROJECTS

BUILD YOUR OWN WORKING

FIBEROPTIC INFRARED AND LASER SPACE-AGE PROJECTS

ROBERT E. IANNINI

TAB BOOKS
Blue Ridge Summit, PA

FIRST EDITION
SEVENTH PRINTING

© 1987 by **TAB Books**.
TAB Books is a division of McGraw-Hill, Inc.

Printed in the United States of America. All rights reserved. The publisher takes no
responsibility for the use of any of the materials or methods described in this book,
nor for the products thereof.

Library of Congress Cataloging-in-Publication Data

Iannini, Robert E.
    Build your own working fiberoptic, infrared, and
laser space-age projects.

    Includes index.
    1. Lasers—Amateurs' manuals.   I. Title.
TA1680.I18   1987   621.366   86-14554
ISBN 0-8306-0824-9
ISBN 0-8306-2724-3 (pbk.)

Front cover photographs: top right, photograph by Robert E. Iannini, middle right
photograph courtesy of Wayne Strattman of Strattman Design, bottom right
photograph by Steven Tidd, and bottom left photograph by Dan Perkins, Las Vegas,
Nevada.

Contents

Acknowledgments

I WISH TO THANK THE ENTIRE STAFF OF INFORMATION UNLIMITED FOR THEIR PA-tience, understanding, and contributions that made this book possible. Special thanks are to go to my layout man and shop technical advisor Richard Upham for his contribution to the mechanical design and layout of the described devices. And to Janet Vandeberghe who typed the text for this book along with the many changes, revisions, and corrections that constantly plagued her efforts. Also the excellent reference and descriptive material obtained from the Hamamatsu, EGG, Ma-Com Laser Diode, and ITT Company has been very valuable.

BUILD YOUR OWN WORKING
FIBEROPTIC
INFRARED AND LASER
SPACE-AGE PROJECTS

Introduction

N EVER BEFORE SINCE THE ADVENT OF THE NUCLEAR ENERGY AGE HAS ANY SCIentific discovery taken such precedence as the laser. The word itself lends an air of mysticism and is often used in countless advertising and attention-getting displays and schemes. It is an acronym for Light Amplification by Stimulated Emission of Radiation. Lasers have a wide range of usage, from destructive beams capable of destroying enemy missiles such as those under consideration for the SDI program to small miniscule powered devices for extracting the information on laser mini-disc players. As a surgical tool the laser far surpasses the knife and scalpel for precision and accuracy. A possible answer to the energy situation suggests the use of these devices for initiating a controlled thermonuclear reaction. The laser like the semiconductor has proven to be of enormous benefit to all fields of science, technology, and everyday life. It is hard to find a lifestyle that is unaffected by this space-age device.

Laser light unlike conventional light has several properties that make it unique. Conventional light such as that from the sun, electric lights, fire, etc., produce photons of energy by spontaneous emission. This energy exists in a wide band or frequency spectrum and occurs as the result of randomly spontaneous energy transitions. These occurrences are produced at different spatial points in the active medium causing deterioration in the coherence quality of the light. The light now contains a wide bandwidth and spreads rapidly due to these properties. Laser light on the other hand has high temporal and spatial coherence that allows it to be projected to great distances with a minimum of spreading. The energy density at a distance now becomes equivalent to many times that of a similar conventional spontaneous source at an equal distance. One now sees the distinct advantage using laser energy as a destructive weapon for actually destroying a target. History mentions when the shields of many soldiers were highly polished, each concentrated incoherent sunlight on the sails of marauding ships and catching them on fire. Fact or fiction, the use of light as a weapon has been considered for many years. The Buck Rogers ray gun seen in the movies in the 1930s has long been sought for duplication as the ultimate weapon of the future. Now it is no longer fantasy.

Developments in laser technology along these lines shows a high feasibility in not only using the devices against missiles and aircraft but also as highly effective antipersonnel weapons. On the other side of the scale these devices have opened up new means of communications where glass fibers (fiberoptics) are used to transmit thousands of times more information using modulated laser beams than is possible with copper wires. Applications in the metal working industry, medical, special effects, and many other fields have made lasers "a major topic of the 20th century."

This book describes in detail several laser projects that are capable of being constructed and used in many practical applications. These projects range from simple milliwatt-powered optical lasers to a multiwatt burning and cutting device. Unfortunately, lasers require high-voltage power supplies very similar to ham radio transmitters and consequently must be treated with total respect for all safety precautions. It is suggested that the hobbyist obtain *The Radio Amateur's Handbook* and study it. This excellent reference shows construction practices and procedures and many similar circuits that are used for laser power supplies.

This book also contains plans for many other useful scientific and electronic projects such as high-voltage power supplies, plasma-producing devices, and laboratory and science fair projects. All information including sources of individual parts as well as complete kits are provided so the builder may construct a useful working device. Reference is given throughout the text to the author's company, Information Unlimited, Inc., as a source of these parts and the builder may also call or write for assistance if needed.

Laser Safety and
General Precautions

MOST LASER PROJECTS THAT INVOLVE SUFFICIENT POWER LEVELS TO BURN OR cut usually require power supplies capable of producing severe electrical shock hazards. It must be stressed to the builder to exercise necessary safety precautions involving proper handling, building, and labeling of potential danger spots. Some of the power supplies used here are very similar to those used in ham radio transmitters; therefore it is suggested that the builder consult *The Radio Amateur's Handbook* on safety and construction practices.

High-powered lasers of the Class IV type such as those described in Chapters 11 and 12 will generate a beam of energy that can start fires, burn flesh, ignite combustible materials and cause permanent eye damage from even unexpected reflections. We therefore present the following section dedicated to proper labeling and the necessary safety requirements as specified by the Bureau of Radiological Health and Welfare. See Tables I-1 and I-2. Also see Table I-3.

Table I-1. Tabulation Chart of FDA Requirements for Laser Products.

Requirements	Class[1]					
	I	IIa	II	IIIa	IIIb	IV
Performance (all laser products)						
Protective housing (1040.10(f)(1))	R^2	R^2	R^2	R^2	R^2	R^2
Safety interlock (1040.10(f)(2))	$R^{3,4}$	$R^{3,4}$	$R^{3,4}$	$R^{3,4}$	$R^{3,4}$	$R^{3,4}$
Location of controls (1040.10(f)(7))	N/A	R	R	R	R	R
Viewing optics (1040.10(f)(8))	R	R	R	R	R	R
Scanning safeguard (1040.10(f)(9))	R	R	R	R	R	R
Performance (laser systems)						
Remote control connector(1040.10(f)(3))	N/A	N/A	N/A	N/A	R	R
Key control (1040.10(f)(4))	N/A	N/A	N/A	N/A	R	R
Emission indicator (1040.10(f)(5))	N/A	N/A	R	R	R^{10}	R^{10}
Beam attenuator (1040.10(f)(6))	N/A	N?A	R	R	R	R
Reset (1040.10(f)(10))	N/A	N/A	N/A	N/A	N/A	R^{13}
Performance (specific purpose products)						
Medical (1040.11(a))	S	S	S	S^8	S^8	S^8
Surveying, leveling, alignment (1040.11(b))	S	S	S	S	NP	NP
Demonstration (1040.11(c))	S	S	S	S	S^{11}	S^{11}
Labeling (all laser products)						
Certification & identification (1010.2, .3)	R	R	R	R	R	R
Protective housing (1040.10(g)(6),(7))	D^5	R^5	R^5	R^5	R^5	R^5
Aperture (1040.10(g)(4))	N/A	N/A	R	R	R	R
Class warning (1040.10(g)(1),(2),(3))	N/A	R^6	R^7	R^9	R^{12}	R^{12}
Information (all laser products)						
User information (1040.10(h)(1))	R	R	R	R	R	R
Product literature (1040.10(h)(2)(i))	N/A	R	R	R	R	R
Service information (1040.10(h)(2)(ii))	R	R	R	R	R	R

Abbreviations

R   -   Required.

N/A -   Not applicable.

S   -   Requirements: Same as for other products of that Class. Also see footnotes.

NP -   Not permitted.

D   -   Depends on level of interior radiation.

Footnotes

[1] Based on highest level accessible during operation.

[2] Required wherever & whenever human access to laser radiation above Class I limits is not needed for product to perform its function.

[3] Required for protective housings opened during operation or maintenance, if human access thus gained is not always necessary when housing is open.

[4] Interlock requirements vary according to Class of internal radiation.

[5] Wording depends on level & wavelength of laser radiation within protective housing.

[6] Warning statement label.

[7] CAUTION logotype.

[8] Requires means to measure level of laser radiation intended to irradiate the body.

[9] CAUTION if 2.5 mW cm^{-2} or less, DANGER if greater than 2.5 mW cm^{-2}.

[10] Delay required between indication & emission.

[11] Variance required for Class IIIb or IV demonstration laser products and light shows.

[12] DANGER logotype.

[13] Required after August 20, 1986.

Table I-2. Laser Labelling and Compliances as Required by the Bureau of Radiological Health.

Label	Warning	Product	Aperture Label	Chapter Reference
CAUTION LASER RADIATION — DO NOT STARE INTO BEAM Helium Neon Laser 1 milliwatt max CW		CLASS II LASER PRODUCT	AVOID EXPOSURE LASER RADIATION IS EMITTED FORM THIS APERATURE	Chapter 10 for 1 mW version
DANGER LASER RADIATION – AVOID DIRECT EYE EXPOSURE HeNe 5 mW max		CLASS IIIA LASER PRODUCT	AVOID EXPOSURE LASER RADIATION IS EMITTED FORM THIS APERATURE	Chapter 6 & 10 for up to 5 mW laser tube
DANGER LASER RADIATION – AVOID DIRECT EXPOSURE TO BEAM MAX PEAK Power 40 W wavelength 904 nm		CLASS IIIb LASER PRODUCT	AVOID EXPOSURE INVISIBLE LASER RADIATION IS EMITTED FROM THIS APERATURE DANGER LASER RADIATION WHEN OPEN AVOID DIRECT EXPOSURE TO BEAM	Chapter 1 up to 40 watts pulsed
DANGER LASER RADIATION – AVOID DIRECT EXPOSURE TO BEAM MAX Power of the wavelength 630 nm		CLASS IIIb LASER PRODUCT	AVOID EXPOSURE INVISIBLE LASER RADIATION IS EMITTED FORM THIS APERATURE	Chapter 4 up to 15 mW continuous
DANGER INVISIBLE LASER RADIATION – AVOID EYE OR SKIN EXPOSURE TO DIRECT OR SCATTERED RADIATION 				

Chapter 12 40 watts 10.6 microns

AVOID EXPOSURE
INVISIBLE LASER RADIATION
IS EMITTED FROM THIS APERATURE

40 watt CO₂
10.6 microns

CLASS IV LASER PRODUCT

DANGER
INVISIBLE RADIATION WHEN OPEN
AVOID DIRECT EXPOSURE TO BEAM

DANGER

HIGH
VOLTAGE

DANGER
LASER RADIATION –
AVOID EYE OR SKIN EXPOSURE
TO DIRECT OR SCATTERED
RADIATION

3 JOULE RUBY
CLASS IV LASER PRODUCT

DANGER

AVOID EXPOSURE
LASER RADIATION IS EMITTED
FORM THIS APERATURE

Chapter 11 3 joule 6292Å

DANGER
LASER RADIATION WHEN OPEN
AVOID DIRECT EXPOSURE TO BEAM

DANGER

HIGH
VOLTAGE

Chapter 14

DANGER
LASER RADIATION WHEN OPEN
AVOID DIRECT EXPOSURE TO BEAM

DANGER

HIGH
VOLTAGE

Certification Label Required on Laser Products

Manufactured By:
INFORMATION unlimited
BOX 716   AMHERST, N.H. 03031

Model Number   X X X X

Serial Number   X X X X X

Manufactured Date   X X X X

This Laser Product conforms to the provisions
of 21 CFR 1040.10 and 1040.11.
Made in U.S.A.

Table I-3. Major Categories of Lasers. (Continued through page xx.)

NAME	WAVE LENGTH	PWR LIMITS	MODE	LASANT	EXCITATION	HIGHLIGHTS	OTHER
RUBY	6934 Å	Megawatt pulses	pulsed & Q switched	Al_2O_3 and Chromium solid rod	Optical pulses from flash lamp	This laser started it all! High peak power, low efficiency visible red. Easily Q switched for high powered optical pulses.	Easy to build Moderately expensive
Nd-YAG	1.06 μm	Megawatt pulses 5 kW cont	Pulsed Q switch continuous	Yttrium Aluminum Garnet solid rod	Flash lamps Arc lamps	Higher efficient than its ruby counter parts. Lowest threshold for lasing. Can be excited continuously with arc lamps if properly cooled.	Easy to build Rod is expensive
Nd	1.06 μm	Multi megawatt pulses	Pulsed Q switch	Neodymium glass solid rod	Flash lamps	Low rep rate-extremely high powered pulses-excellent for Q switching producing terawatts of peak power - candidate for fusion.	Easy to build Moderately expensive
LIQUID DYE	Multi wavelength usually tunable	Megawatt pulses	Pulsed Q switch	Dye or other solution (chelates)	Flash lamps or other laser	Tunable devices use various liquids with fluorescent qualities. Produce short high powered pulses. Good R&D studies. Tunable continuous operation possible.	Hard to build Expensive
DIODE LASERS	1.06 μm experimental visible	1 to 100 watt pulsed 1 mW to 1 watt continuous	Pulsed continuous	GaAs and other exotics	Electrical pulse or DC for continuous	Rugged, compact, reliable used for communications, consumer products with fibre optics. Little support equipment required, easy to detect. High powered output with multi-diode arrays.	Easy to build

Type	Wavelength	Power	Operation	Medium	Excitation	Applications	Notes
CHEMICAL	2.5 to 4 μm	Up to 50 megawatts	Continuous for short periods	Hf Df	Self excited by chemical reaction	Weapons type application utilize direct chemical reaction energy. Beam intensities of 50 million watt occurring for 10-20 secs duration can destroy most objects in its path. Candidate for SDI.	Classified Dangerous Impractical to build
FREE ELECTRON	Multi wavelength visible, UV, X	Megawatt beam of energy	Continuous	Electron beam in spatially changing magnetic field		This is the weapon we have all been waiting for. Unavailable to the public, highly classified with continuous beam power up to 1 mW in the UV range. Obvious candidate for SDI and any military weapon application. Needs a lot of support equipment but is the most promising.	Classified Impossible to build due to parameters
ARGON	4880, 5140 Å	.5 mW to 100 watts	Pulsed or continuous	Argon & Helium Gases	Electrical Discharge Other Lasers	Special effects, holography, surgery, light shows	Hard to build Expensive
KRPTON ION	Other in Red/Yellow						
HELIUM-NEON	6328 Å	.1 mW to 100 mW	Continuous	Helium Neon Gases	Electrical Discharge	General purpose, highly popular for just about all applications not involving burning or cutting. Excellent beam quality & coherent properties Low cost consumer usage. Green line available.	East to build popular. Not excessively dangerous
HELIUM-CADMIUM	4416 Å 3250 Å	.1 mW to 100 mW	Continuous	Helium & Cadmium Vapor	Electrical Discharge	Shortest wavelength continuous, commercially available device.	Hard to build Expensive

	Wavelength	Power	Mode	Medium	Excitation	Comments	Build
NITROGEN	3370 Å	1 to 100 kW	Short pulses	Nitrogen gas	Electrical pulses	High powered extremely short pulses of ultraviolet. Many R&D applications involving fluorescent studies, etc. High gain lasing seldom even requires optics.	Easy to build
EXCIMER	Ultraviolet	High kW	Short Pulses	Ar Fl, Kr Fl, Xe Cl, Ye Br	Electrical	Extremely high powered short pulses in the UV range. Utilizes a diatomic molecule that only exists in its excited state.	Hard to build
IODINE	1.3 μm	High kW	Pulsed Continuous	Iodine Oxygen	Optical Chemical	High powered laser produces 1.3 μm by photo dissociation. Possible candidate for weapons.	Classified not feasible to build
COPPER	5106 Å, 5782 Å	1 to 100 W	Pulsed	Copper vapor	Electrical pulses	Fairly new laser is much more efficient than the argon ion-can also use gold vapor.	Hard to build but possible for those advanced
CARBON DIOXIDE	10.6 μm	1 to high kW	Pulsed/ continuous	CO_2 N_2 H_2 Gases	Electrical dc, ac, rf	Highly efficient, high gain work horse. Is the industry standard for metal working, materials processing or wherever high power is required. Considered for weapon use, fusion activation.	Easy to build Classified Version for weaponery
X-RAY	Multi wavelength UV to X-ray	Multi megawatt		Certain metals exposed to a nuclear reaction		Controversial weapon-great for SDI but requires a nuclear excitation. Multiport device can direct powerful short beams of UV/X-ray many 1000 of kilometers in space. Impractical for terrestial use due to absorption.	Highly Classified Hush Hush Hobbyists are advised not to build this device in a crowded neighborhood

Project Descriptions and Special Precautions

ALL OF THE NECESSARY MATERIALS FOR COMPLETING THE PROJECTS IN THIS book are available from Information Unlimited, P.O. Box 716, Amherst, NH 03031. Write or call (603) 673-4730.

SSL3—Solid-State Gallium-Arsenide Injection Laser System. This useful lab device produces 4- to 30-watt peak power infrared pulses at 200 to 2500 pulses per second repetition rate. It is labeled Class IIIB and requires other support equipment such as the laser receiver LSD3 described in Project 2 to be useful. This is a good advanced science fair project for high school. Cost of completion is approximately $100.00.

LSD3—Laser Light Detector. This is a high sensitivity, fast response laser receiver capable of detecting pulses from the SSL3 described in Project 1. Unit is shown with basic optics and reproduces the pulse received. A special output that produces an expanded output is shown. Cost of completion is about $75.00.

SD5—Infrared Viewer. Excellent device with many useful functions ranging from night time surveillance to viewing IR laser beams such as that shown in Project 1. It has many uses and is an excellent science fair project. Cost of completion is about $150.00.

CWL1—Continuous-Wave Solid-State Laser. This project shows a simple safe way of powering a cw laser diode. Good for use with fiberoptics or general lab work. As of now these laser devices are very expensive and extremely prone to being destroyed. They are more for educational use than anything else with a cost of over $100.00 for the laser diode.

Fiberoptics. This project shows a circut capable of being used for both receiving and transmitting over an optical fiber. Several references are made to available kits and other paraphernalia. A complete and extensive introduction to fiberoptics for communications is included.

HNE3—Helium-Neon Visible-Red Laser. This project shows how to construct a very versatile power supply for powering different helium-neon laser tubes from 0.3 to 5 mW. Cost is approximately $100.00. Excellent lab or evaluation device for testing various types of laser tubes.

HNM1—Voice Modulation of a Helium-Neon Laser. This project shows low to modulate a laser beam and actually voice communicate. Works well with the optical voice receiver in Chapter 8. Excellent science fair project, requires a helium-neon laser system with special ballasting. Cost is approximately $300.00.

LLD3—Optical Light Detector and Voice Receiver. This device is capable of listening to any varying periodic light source. To be used as a receiver to the HNM1. Excellent science fair project costing approximately $50.00.

LDT1—Optical Switch. This simple circuit allows controlling various electrical devices via an optical link. Great for shooting galleries, alarms, counters, etc. Very useful for practicing with gun sighting. Cost is about $30.00.

LGU6—Visible-Red Hand-Held Laser Light Gun. This project shows how to build several different versions of this extremely popular laser. All self-contained lightweight and efficient circuitry provides a very useful and interesting project. Cost is $100.00 to $200.00 depending on power level. Safe to use and great science fair project. Use with the LDT1 described in Chapter 9.

RUB5—Ruby Laser Gun. This is an advanced project that has proven very popular. It is costly and involves high voltages and high optical output capable of punching holes in the hardest of metals. It can be built utilizing neodymium glass, yag or ruby as the active lasting medium and will cost accordingly.

LC7—Carbon Dioxide Laser. This laser is capable of producing a continuous 30- to 50-watts of energy that can fabricate plastics, textiles, light metal, or etch wood and a list of other similar feats. This is a revised system that was shown in my first book, *Build Your Own Laser, Phaser, Ion Ray Gun and Other Working Space-Age Projects* (TAB book #1604). It would cost thousands of dollars to purchase a laser with the same capabilities. It is an advanced project costing over $500.00. Caution must be exercised with the output beam and the high voltages required.

PTG1—Plasma Tornado Generator. This project shows how to construct a visual special effects device capable of demonstrating the weird and bizarre effects of electric plasma. This is a very popular device for the effects produced. The project costs under $100.00 and can be made identical to the round plasma sculpture globes featured in specialty catalogs for hundreds of dollars.

HVG1—High-Voltage Generator. This is a high-voltage laboratory device that is useful in all types of lasers, plasma ion, and particle applications. It also can be used for lightning displays and special effects. Excellent lab device costing $300.00.

Chapter 1

LASER RADIATION –
AVOID DIRECT EXPOSURE
TO BEAM

MAX PEAK POWER 40 W
WAVELENGTH 904 nm

CLASS IIIb LASER PRODUCT

Solid-State
Laboratory IR Laser (SSL3)

T HIS PROJECT SHOWS HOW TO CONSTRUCT A GALLIUM ARSENIDE INJECTION LA-
ser providing a medium powered class IIIB source of optical IR laser energy. The
device is shown built in two sections: the first the "Power Supply and Pulse Driver
Circuit" and the second the "Laser Head Optic Section." These are connected via an
umbilical cable up to a length of 10 feet or more. Remote control connections are available from the power supply section. The system derives its power from the standard
115-volt ac lines through a polarized plug. Power requirements are minimal being around
20 watts.

The unit is intended for laboratory and experimental use and is capable of producing 50 to 2,000 pps of 50 watt, 200 nanosec optical power at 9000 Å. Regulation is
less than 3% throughout the pulse range. The head section can easily be mounted and
secured for interfacing with other optical systems.

The device generates an adjustable frequency of low- to medium-
powered IR pulses of invisible energy and must be treated with care. At
no time should it be pointed at anyone or anything that could reflect these pulses. Never
look into the unit when the power is on. It is intended to be used for ranging, simulated
weapons practice, intrusion detection, communications and signaling along with a variety of related scientific, optical experiments and users.

THEORY OF OPERATION

A laser diode is nothing more than a three-layer device consisting of a pn-junction
of n-type silicon, a p-type of gallium arsenide and a third p-layer of doped *gallium arsenide* with aluminum.

The n-type material contains electrons that readily migrate across the pn-junction and fill the holes of the p-type material, conversely holes in the p-type migrate to the n-type material and join with electrons. This migration causes a potential hill or barrier consisting of negative charges in the p-type material and positive charges in the n-type material that eventually ceases growing when a charge equilibrium exists. In order for current to flow in this device, it must be supplied at a voltage to overcome this potential barrier. This is the forward voltage drop across a common diode. If this voltage polarity is reversed, the potential barrier is simply increased assuring no current flow. This is the reversed bias condition of a common diode.

A diode with no external voltage applied to it contains electrons that move and wander through the lattice structure at a low, lazy average velocity as a function of temperature. When an external current at a voltage exceeding barrier potential is applied, these lazy electrons now increase their velocity to where some by colliding acquire a discrete amount of energy and become unstable eventually emitting this acquired energy in the form of a photon upon returning to a lower energy state. These photons of energy are random both in time and direction, hence any radiation produced is incoherent such as that of a light-emitting diode.

The requirements for coherent radiation is that these discrete packets of radiation be in the form of lockstep phase and in a definite direction. The above demands two essential requirements (A) sufficient electrons at the necessary excited energy levels and (B) an optical resonant cavity capable of trapping these energized electrons for stimulating more and directing them. The number of energized electrons are determined by the forward diode current. A definite threshold condition exists where the device emits laser light rather than incoherent as a light-emitting diode. This is why the device must be pulsed with high current. The radiation from these energized electrons is reflected back and forth between the square cut edges of the crystal which form reflecting surfaces due to the index of refraction of the material and air. The electrons are initially energized in the region of the pn-junction. When these energized electrons drift into the p-type transparent region, they spontaneously liberate other photons that travel back and forth in the optical cavity interacting with other electrons commencing laser action. A portion of the radiation traveling back and forth between the reflecting surfaces of these mirrors escapes and constitutes the output of the device.

POWER SUPPLY

Ac power is obtained via polarized plug (CO1) through fuse safety resistors (R1) and (R2). These are low-ohm, low-watt film resistors that will quickly open up in case of a gross circuit fault. Proper polarization in respect to the ac lines is a necessity to prevent unnecessary shock hazard and use of grounded test and measuring equipment. (S1) is a key switch where the key can only be removed in the "off" position. The power indicator lamp consists of neon lamp (NE1) and associated current-limiting resistor (R3).

Diode (D1) and (D2) along with capacitors (C1) and (C2) comprise a voltage doubler eliminating the usual iron and copper transformer. The voltage across C1 and C2 is 1.4×230 or approximately 340 volts. Zener diodes (Z1) through (Z5) are selected to provide the proper required current pulse for the laser diode in operation. Each one of these diodes provides a 15-volt drop. The builder may wish to provide a multi-position switch to select these various combinations of zener diodes externally via a front panel control. This allows final tailoring of the laser current pulse.

The trigger circuit consisting of unijunction (Q2) derives its power through R4 and zener diode (Z6). A delay equal to the time required for C3 to charge to its zenering voltage is provided by these components. This voltage is regulated to 15 volts via zener diode (Z7). It should be noted that the actual delay time is considerably less than the RC value of R4 and C3 and hence must be taken into account when selecting C3. Remote control is via an external jumper connection in series with the trigger output pulse to SCR1. The umbilical cable now interconnects the two sections.

TRIGGER CIRCUIT

This circuit determines the pulse repetition rate of the laser and uses a unijunction (Q2) whose pulse rate is determined by resistor (R7) and capacitor (C4). You will note that the maximum permissible pulse rate is determined by the laser diode rating, the RC time constant of the charging circuit and the current capability of the power supply. R7 is controlled via the front panel. Resistor (R8) sets the upper pulse rate limit.

DISCHARGE CIRCUIT

The discharge circuit generates the current pulse in the laser and consequently is the most important section of the pulser. The basic configuration of the pulse power supply is shown in the system schematic. The current pulse is generated by the charging storage capacitor (C5) through SCR and laser diode (D5). The rise time of the current pulse is usually determined by Q3 while the fall time is determined by the capacitor value and the total resistance in the discharge circuit. Table 1-1A shows typical anode voltage and current wave forms of the SCR during the current pulse through the diode laser.

The peak current, pulse width and voltage of the capacitor discharge circuit are interrelated for various load and capacitance values. The peak laser current and charged capacitor voltage relationships are given in Table 1-1B for several different capacitor values and typical laser types. The voltage and current limits of the SCR are also shown.

Short pulse widths provide less time for the SCR to turn on than longer pulse widths; therefore, the SCR impedance is higher and more voltage is required to generate the same current. Table 1-1C shows the current pulse waveforms for the three different values of the capacitance. The capacitor is charged to the same voltage in all three cases, i.e., approximately 200 volts.

SCR SWITCH

In conventional operation of an SCR, the anode current, initiated by a gate pulse, rises to its maximum value in about 1 microsecond. During this time the anode-to-cathode impedance drops from open circuit to a fraction of an ohm. In injection laser pulsers, however, the duration of the anode-cathode pulse is much less than the time required for the SCR to turn on completely. Therefore, the anode-to-cathode impedance is at the level of 1 to 10 ohms throughout most of the conduction period. The major disadvantage of the high SCR impedance is that it causes low circuit efficiency. For example, at a current of 40 amps, maximum voltage would be across the SCR while only 9 volts would be across LA1. These values represent a very low circuit efficiency.

Table 1-1. Graphs and Charts

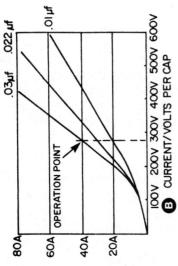

Ⓐ TYPICAL SCR VOLTAGE/PEAK CURRENT

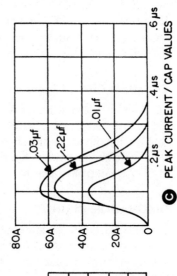

Ⓑ CURRENT/VOLTS PER CAP

TYPICAL CURVES AND PARAMETERS FOR LASER
PULSERS NOT NECESSARILY THOSE OBSERVED
IN THIS CIRCUIT

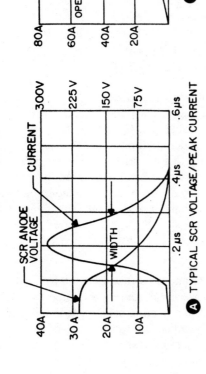

Ⓒ PEAK CURRENT / CAP VALUES

WAVELENGTH-9040 Å
PULSE DURATION-200 NANO SECS

LASER	POWER	Ipk	Cx	ZENER	PULSE/SEC
LD74	5-7 WATTS	40 AMPS	.033	90V	2000
LD66	8-10 WATTS	40 AMPS	.033	90V	2000
LD1630	25-30WATTS	40 AMPS	.033	90V	2000
LD780	20-25WATTS	40 AMPS	.033	90V	2000

Ⓓ LASER SELECTION CHART

CHART FOR EXPERMENTAL
RESULTS USING SELECTED
VALUES OF CX AND RX.

The efficiency of a laser "array" is greater due to its circuit impedance being more significant.

Because the SCR is used unconventionally, many of the standard specifications such as peak-current reverse-voltage, on-state forward voltage, and turn-off time are not applicable. In fact, it is difficult to select an SCR for a pulsing circuit on the basis of normally specified characteristics. The specifications important to laser pulser applications are forward-blocking-voltage and current-rise-time. A "Use Test" is the best and many times, the only practical method of determining the suitability of a particular SCR.

STORAGE CAPACITOR

The voltage rating of the storage capacitor (C5) must be at least as high as the supply voltage. With the exception of ceramic types, most capacitors (metallized paper, mica, etc) will perform well in this circuit. Ceramic capacitors have noticeably greater series resistance, but are usable in slower speed pulsing circuits.

LAYOUT WIRING

Lead lengths and circuit layout are very important to the performance of the discharge circuit. Lead inductance affects the rise time and peak value of the current and can also produce ringing and an undershoot in the current waveform that can destroy the laser.

A well built discharge circuit might have a total lead length of only one inch and therefore an inductance of approximately 20 nanohenries. If the current rises to 75 amperes in 100 nanoseconds, the inductive voltage drop will be e = L di/dt. If proper care is not taken in wiring the discharge circuits, high inductance voltage drops will result.

A one-ohm resistor in the discharge circuit will greatly reduce the current undershoot in single diode lasers. Laser "arrays" usually have sufficient resistance to eliminate undershoot. The small resistance in the discharge circuit is also useful in monitoring the laser current, as described in the following section.

A clamping diode (D4) is added in parallel with the laser to reduce the current undershoot. Its polarity should be opposite to that of the laser. Although the clamping diode is operated above its usual maximum current rating, the current undershoot caused by ringing is very short and the operating life of the diode is satisfactory.

CURRENT MONITOR

The current monitor in the discharge circuit provides a means of observing the laser current waveform with an oscilloscope. A resistive type monitor (R11), reduces circuit ringing and current undershoot, but the lead inductance of the resistor may cause a higher than actual current reading. A current transformer such as the Tektronix CT-2 can also be used to monitor the current and is not affected by lead inductance. Because the transformer does not respond to low-frequency signals, it should be used with fast time, short pulse-width, fast fall-time waveforms.

CHARGING CIRCUITS

The second major section of the pulser is the charging circuit. The circuit charges the capacitor to the supply voltage during the time interval between laser current pulses, and isolates the supply voltage from the discharge circuit during the laser current pulse, thereby allowing the SCR to recover the blocking state. Because the response times of the charging circuit are relatively long, lead lengths are not important and the circuit can be remotely located from the discharge circuit.

The simplest charging circuit is a resistor/cap combination. The resistor must limit the current to a value less than the SCR holding current, but should be as low as practical because this resistance also determines the charging time of the capacitor C5.

For example, a resistance of 40 kilohms limits the current to 10 milliampers from a 400-volt supply. This current value is just at the holding level of an average 2N2443 SCR. A time of almost 3 milliseconds is required to charge a 0.02 microfarad capacitor to the supply voltage in three time constants through a 40-kilohm resistor. Therefore the pulse rep rate (PRR) of the pulsing circuit is limited to about 375 Hz. If the PRR exceeds this value, the capacitor does not completely recharge between pulses and the peak laser current decreases with increasing PRR.

The peak current in the discharge circuit may be controlled by varying the supply voltage, provided the PRR is low enough to allow the capacitor to recharge fully between current pulses. Both the supply voltage and the PRR determines the peak laser current. There is considerable risk in increasing the supply voltage to compensate for insufficient recharge time. If the PRR is decreased while the supply voltage is high, the capacitor again recharges completely and the laser pulse current increases to a value that may damage or destroy the laser diode.

The need for a variable voltage supply and the low PRR limit are the major disadvantages of the resistive type charging network. Therefore, the limitations of the simple resistor drivers is that a resistor large enough to keep the current below the holding current of the SCR also limits the pulse repetition rate. The frequency capability of the pulse power supply can be improved with the charging circuits used in these plans. Capacitor C5 is charged to the supply voltage when the SCR is not conducting. Diodes D3 and D6 are also in the off state, because zero current flows through them.

At the onset of a pulse trigger, the impedance of the SCR drops rapidly and capacitor C5 discharges toward ground potential. During the period when current is surging through diodes D3 and D6, transistor Q1 is reverse biased by the diode voltage drop. The SCR turns off when the current drops below its holding current, which determines the value of R5. The voltage across capacitor C5 then charges back to the supply voltage. During the capacitor charge cycle diodes D3 and D6 pass no current and Q1 is forward biased into the saturation region. Obviously, the charging time of C5 is now shorter due to being charged through the small resistor R6.

CONSTRUCTION STEPS

1. Layout and identify all parts and pieces. Separate the components that go to the "power supply" assembly board and the "laser head" assembly board. See Table 1-2 and Fig. 1-1.

2. Fabricate perfboard PB1 from a 2.8 × 2.8 piece of .1 × .1 grid. Note the border holes shown in Fig. 1-2A.

Table 1-2. Parts List SSL3.

R1,2	2	10-ohm 1/4-watt resistor
R3	1	100 kΩ 1/4-watt resistor
R4	2	Use (2) 20 kΩ 5-watt resistors in series for 40 kΩ at 10 watts
R5	1	33 kΩ 1-watt resistor
R6	4	1.8 kΩ 3-watt in series/parallel or use single 1.8 kΩ 10-watt resistor
R7	1	100 kΩ Pot resistor
R8	1	25 kΩ trimpot resistor
R9,10	2	100-ohm 1/4-watt resistor
R11	1	1-ohm 1/2-watt carbon comp resistor
C1,C2	2	33 μF/250 V elect
C3	1	220 μF/50 V elect (voltage rating important)
C4	1	.05 μF/50 V disc
C5	1	.033 μF/400 V special discharge capacitor
D1,2,4,5, 7	5	1N4007 diodes 1000 V
D3,6	2	1N914 signal diodes
Z1,2,3,4, 5,6,7,8	8	1N5245 15 V zener diodes
Q1	1	2N3439 high volt NPN
Q2	1	2N2646 UJT
SCR1	1	Special selected SCR 2N4443
NE1		Neon lamp leads
CO1		3 wire line cord
S1		Key switch (nonremovable when on)
S2		Optional multiposition switch for control of pulse current
J1		RCA phono jack
WR1	24 inches	#24 vinyl jacket hook up wire
WR2	24 inches	#20 vinyl jacket hook up wire
WR3	4 feet	3 cond #18 shielded cable (umbilical)
PB1		2.8 inch × 2.8 inch .1 perfboard
PB2		1.5 inch × .8 inch .1 perfboard
BU1,2,5 6	4	3/8 plastic bushing
BU4,3	2	Clamp bushing
CA1,2	2	33 mm brass bottle cap Fig. 1-3
CA3	1	38 mm brass bottle cap
CA4	1	1-5/8 inch plastic cap
CA5	1	1-1/2 inch plastic plug #988
PL1	1	3-1/2 inch × 3/4 inch .025 inch copper fab Fig. 1-3
EN1	1	3-1/2 inch × 3-1/4 × 2 inch metal cabinet
MK1	1	Mica insulating washer
WN1	1	Wire nut small
KN1	1	Knob
TU1	1	1-1/2 inch × 6 inch × .035 inch plastic tube
TU2	1	1-5/8 × 4-1/2 × .058 plastic tubing
LE1	1	29 × 43 mm lens
SW1,2	2	6-32 × 3/8 nylon screw
NU1	1	6-32 nut
CH1	1	6 inch chain
LAB1	1	Class IIIB warning
LAB2	1	Certification label
LAB3	1	Interlock label
LAB4	1	Aperture label

SSL3K - Kit of Above (Select Diode Below)
SSL30 - Assembled with LD740 Laser Diode

*LA1 - Laser Diodes (Select for desired power level)
LD740 4-6 WATTS 40 AMP
LD660 10-14 WATTS 40 AMP
LD780 20-25 WATTS 40 AMP
LD1630 30-35 WATTS 40 AMP

Safety Glasses - Glendale Optics #LGS1 Tel 516-921-5800

The high speed ultra sensitive optical detector LSD3 described in Chapter 2 allows controlling and communicating information over considerable distances.

All the above parts and kits are available through Information Unlimited, P.O. Box 716, Amherst, N.H. 03031. Write or call (603) 673-4730.

Asterisk parts are individually available from Information Unlimited, P.O. Box 716, Amherst, N.H. 03031. Write or call (603) 673-4730.

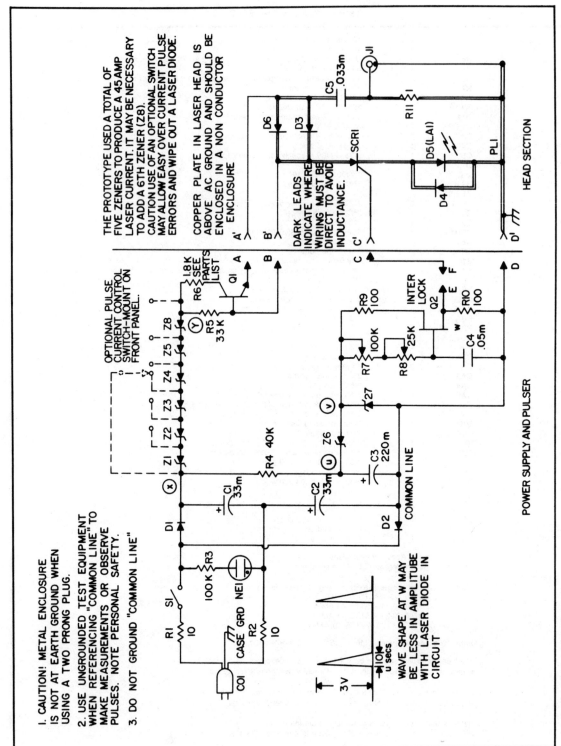

Fig. 1-1. SSL3 IR Laser Schematic.

8

3. Assemble the components as shown in Fig. 1-2A, B, C. Observe polarity of capacitors and semiconductors. Note Fig. 1-2A identifies the components Fig. 1-2B shows the wiring. Use holes shown in perfboard. Fig. 1-3 shows the heat sink.

4. Connect the components using their actual leads wherever possible. Follow the layout and avoid bare wire bridges.

5. Connect wire leads for S1, R7, NE1 and "remote" as shown in Fig. 1-4.

6. Fabricate EN1 as shown Fig. 1-4 inset sketches for dimensions. Note these are not critical, but should allow for clearance and symmetry of mounted components.

7. Mount components as shown and secure.

8. Connect wire leads from board to mounted components (do not secure board in at this time).

9. Obtain a multimeter and an *Ungrounded* oscilloscope and perform the following tests. Place a piece of insulated material such as cardboard between the assembly board and the enclosure to avoid any shorts since it is not yet secured in place.

CAUTION! The "common line" is above earth ground potential. Shock hazard exists for these steps.

A. Connect meter between X and common line. Plug unit into 115 ac (and measure 340 Vdc).

B. Temporarily connect B base of Q1 to common line and measure 15-V drops across each zener diode Z1-Z8. Measure 250 at Y. Voltage at X may not be slightly less than 340. Disconnect B when measurements are completed.

C. Measure 30 Vdc from U to common line.

D. Measure 15 Vdc from V to common line. This is the timing voltage for pulse delay.

E. Connect a *Ungrounded* scope to W and note waveshape as shown in Fig. 1-1. Turn R7 full CW (max rep rate) and set trimpot R8 for 500 μsec separation or 2000 rep per sec. Pulse rate should now be variable between 250 and 2000.

This completes the preliminary test of the Power Supply Section.

Laser Head Section

10. Fabricate PL1, CA1, and CA2 as shown Fig. 1-3. Note when using higher powered LD1630 diode hole will be tapped for 10-32 rather than 8-32 for LD740 and LD660.

11. Layout PB2 as shown in Fig. 1-4A. Use actual holes as shown for components. Pay particular attention to double dashed leads Fig. 1-4B as these must be short and direct as possible to eliminate stray inductance. Note connections to PL1 copper plate. Do not connect LA1 (D5) laser diode at this time.

12. Determine length of interconnecting cable WR3 and connect to laser head assembly board as shown in Fig. 1-4B. Note shield of cable connects directly to PL1. Fab CA5 shown in Fig. 1-5. Connect other end into laser power supply (Fig. 1-6). You may know these for strain relief (not shown).

13. Fabricate TU1 from a 1.5-inch × 6-inch × .035 wall *plastic* tube. Note small hole for SW2 to secure assembly.

14. Fabricate TU2 from a 1.625-inch × 4-1/2-inch × .058 wall plastic tube for telescoping into TU1.

15. Fabricate CA3 lens retainer from a 38mm bottle cap. Remove 1 inch from center

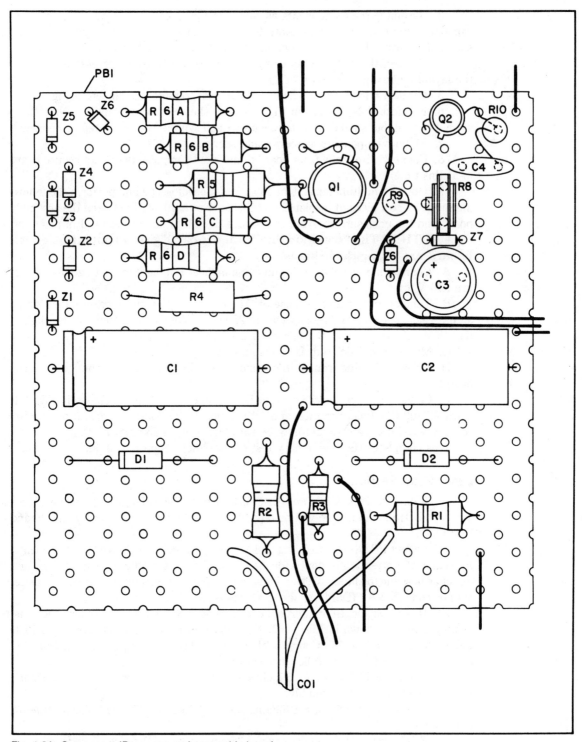

Fig. 1-2A. Component ID power supply assembly board.

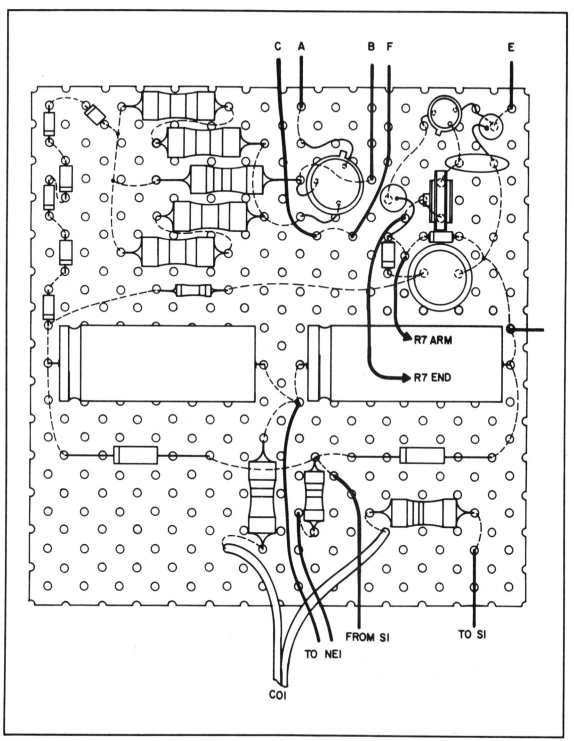

Fig. 1-2B. Wiring aid power supply assembly board top view.

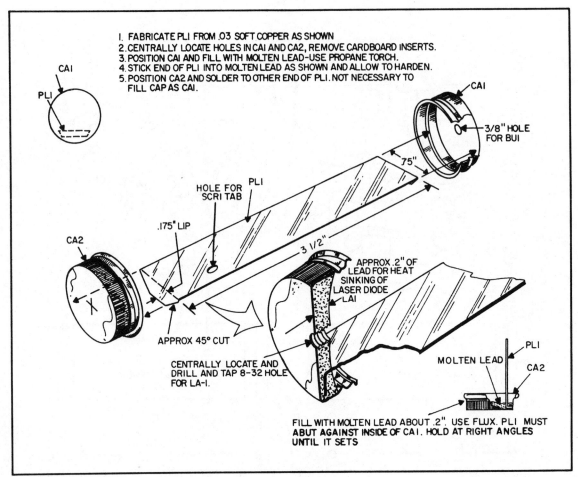

1. FABRICATE PL1 FROM .03 SOFT COPPER AS SHOWN
2. CENTRALLY LOCATE HOLES IN CA1 AND CA2, REMOVE CARDBOARD INSERTS.
3. POSITION CA1 AND FILL WITH MOLTEN LEAD—USE PROPANE TORCH.
4. STICK END OF PL1 INTO MOLTEN LEAD AS SHOWN AND ALLOW TO HARDEN.
5. POSITION CA2 AND SOLDER TO OTHER END OF PL1. NOT NECESSARY TO FILL CAP AS CA1.

Fig. 1-3. Heat sink fab.

for lens aperture. Secure LE1 lens with RTV or other adhesive. Note small hole for retaining chain CH1. Insert into end of TU2 and secure with RTV.

16. Attach other end of chain through the aperture cap CA4 and secure as shown in Fig. 1-7. Chain fits into cap when placed over lens.

17. Label as shown in Figs. 1-8 and 1-9. Note position of labels and correct nomenclature. See Fig. 1-10.

Final Testing

18. Obtain a 1N4007 power diode and temporarily wire it in place of the laser diode LA1. See D5 Fig. 1-5 in dashed lines. The diode simulates the actual laser diodes and avoids costly replacement should a gross error exist. The actual laser diode is not to be wired in at this point.

19. Obtain a scope with a 35 MHz or higher bandwidth. Any scope able to resolve pulses of 0.1 μsec will suffice. **Make certain scope is isolated above earth ground.** Use an adapter plug and doublecheck to prevent case grounding.

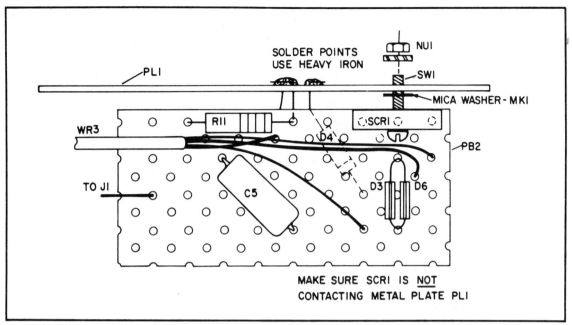

Fig. 1-4A. Component ID laser head assembly board.

20. Connect probe across R11 to copper plate PL1 and observe the waveshape labelled "current" in Table 1-1A. Note this will be inverted from that as shown and will actually be a resultant voltage waveform developed across R11 or a result of the laser current pulse.

21. Pulse amplitude should be no more than 50 volts. It may be necessary to add or subtract zener diodes to tailor this pulse. See Table 1-3D. The range of pulse rep rate should be 200 to 2000 pps. The pulse amplitude should only vary several percent

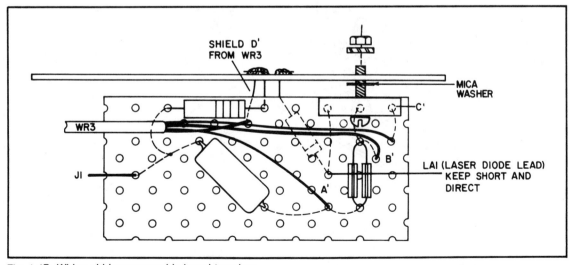

Fig. 1-4B. Wiring aid laser assembly board top view.

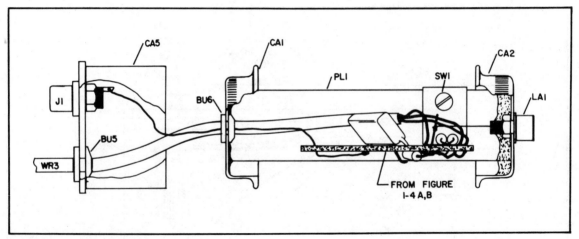

Fig. 1-5. Laser head assembly.

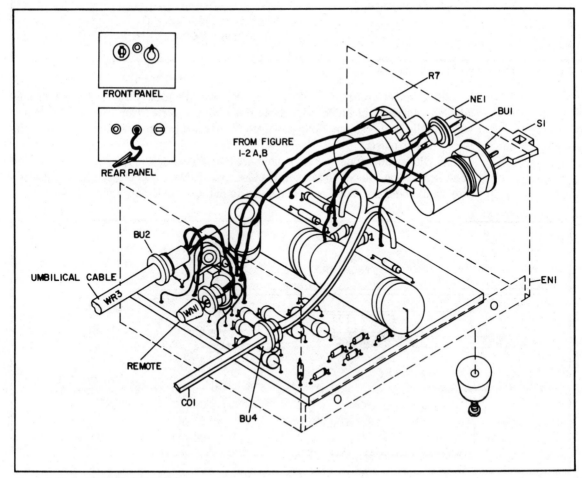

Fig. 1-6. Main assembly internal view.

14

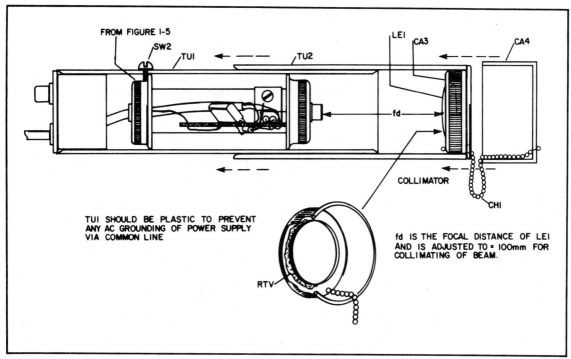

Fig. 1-7. Laser head assembly internal view.

throughout this range. You will note that amplitude will slightly decrease as the rep rate is increased.

At this point you may wish to consider wearing laser safety glasses.

22. Remove the test diode D5 and connect the laser diode. Place an IR indicator

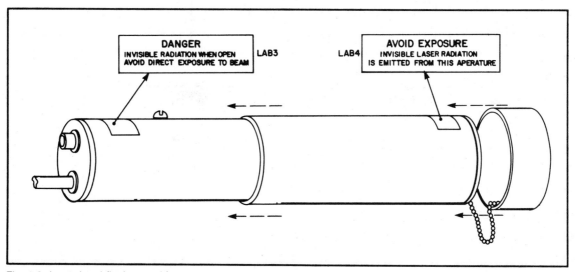

Fig. 1-8. Laser head final assembly.

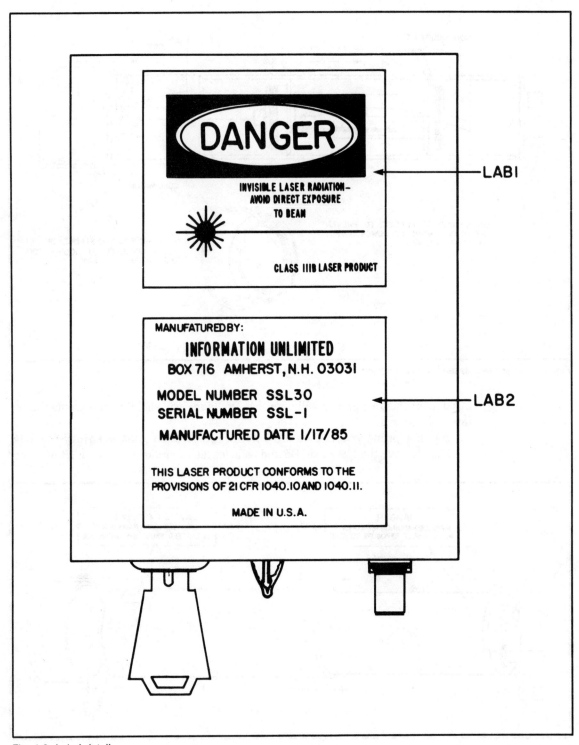

Fig. 1-9. Label detail.

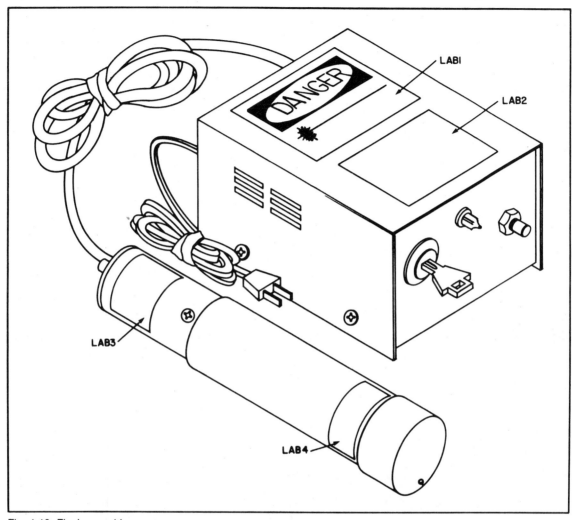

Fig. 1-10. Final assembly.

such as our IRP1 several inches from the diode and note an orange glow when unit is turned on. Recheck pulse shape and amplitude.

23. Power down unit and position laser head assembly (Fig. 1-5) into housing (Fig. 1-7). Secure with nylon screw SW2 as shown. Slide in collimator section and power up unit. Focus to target using our IRP1 IR Indicator or SD5 Night Eye Scope.

Please note that laser diodes are very easily destroyed by over current pulses and over shoot. Pay attention to the correct current pulse in Step 21. Make a final check and fill out Table 1-3.

Table 1-3. Test Report SSL3 IR Laser.

Test Report SSL3 IR Laser.

UNIT _____ For BRH COMPLIANCE ASSEMBLER NAME _____

DATE _____

1. Check action of S1 key switch Fig. 1-1, 1-6.

2. Check action of housing interlocks or use of appropriate labeling for noninterlocked protective housing Fig. 1-8.

3. Test remote control leads Fig. 1-1, 1-6.

4. Check delay action of trigger circuit and igniting of NE1 Fig. 1-1, 1-6.

5. Check for proper labeling of housings Fig. 1-9, 1-10.

6. Check for aperture cap Fig. 1-7, 1-8.

7. Monitor 340 volt at point Y Fig. 1-1.

8. Set pot to lowest pulse rate = 250 (full CCW).

9. Connect frequency meter to SCR gate through 47 kΩ resistor point W.

10. Connect scope across 1-ohm resistor to read 200 nanoseconds pulse of 30-50 amps.

11. Measure the following at the indicated rep rate steps A through E.

Rep Rate	Pulse Amp	Note Volts at Y
2500	38-40 amps	
1000	40-42 amps	
250	42-44 amps	

12. Record and date unit.

Chapter 2

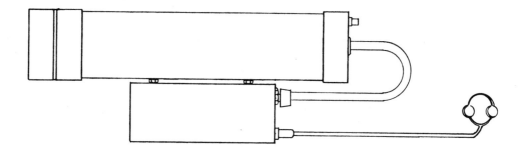

High-Speed
Laser-Light
Pulse Detector (LSD3)

T HIS PROJECT IS A SPECIALIZED DEVICE CAPABLE OF DETECTING HIGH-SPEED pulses. A suitable laser for generating these pulses is the SSL5 described in Chapter 1. The uniqueness of this device is the ability to resolve these pulses with rise times in the nanosecond range. This capability makes it possible to transmit large amounts of information during a small period of time.

The heart of this device is the actual "front end detector," in this case a PIN diode device. The characteristics of this detector are low capacitance, high resistance, and low leakage when operating in the reverse bias mode. It may also be used as a photocell when operated in the forward mode.

Because these devices are such an important tool in the detection of laser pulses we have devoted a section to their theory. We wish to acknowledge the Hamamatsu Company for the following excellent material consisting of both text and illustrations.

THEORY OF SILICON PHOTOCELLS

The following information is courtesy of Hamamatsu Corporation.

Glossary of Terms

Spectral Response
The photocurrent produced by a given level of incident light varies with wavelength. This wavelength/response relationship is known as the spectral response characteristic and is expressed numerically in terms of radiant sensitivity, quantum efficiency, NEP, detectivity, etc.

Radiant Sensitivity
This measure of sensitivity is the ratio of radiation energy expressed in watts incident on the device to the photocurrent output expressed in amperes. It may be expressed as either an absolute sensitivity, i.e., the A/W ratio, or as a relative sensitivity, normalized with respect to the sensitivity at the wavelength of peak sensitivity, with the peak value taken as 100 usually. For the purposes of this catalog, the spectral response range is taken to be the region within which the radiation sensitivity is within 5% of the peak value.

Quantum Efficiency (Q.E.)
This is the ratio of number of incident photons to resulting photoelectrons in the output current, without consideration given to the individual photon energy levels, resulting in a slightly different spectral response characteristic curve from that of the radiant sensitivity.

NEP (Noise Equivalent Power)
This is the amount of light equivalent to the intrinsic noise level of the device. Stated differently, it is the light level required to obtain an S/N ratio of 1. The NEP is one means of expressing the spectral response. In this brochure, the NEP value at the wavelength of maximum response is used. Since the noise level is proportional to the square root of the bandwidth, the NEP is expressed in units of $W/Hz^{1/2}$.

$$NEP = \frac{Noise\ Current\ (A/Hz^{1/2})}{Radiant\ Sensitivity\ at\ Peak\ (A/W)}$$

D* (D-Star)
Detectivity, D, is the inverse of the NEP and is used as a measure of the detection sensitivity of a device. Since noise is normally proportional to the square root of the photosensitive area, the smaller the photosensitive area, the better the apparent NEP and detectivity. To take into consideration material properties, the detectivity D is multiplied by the square root of this area to obtain D*, expressed in units of $cm \cdot Hz^{1/2}/W$. As with NEP, the values used herein are those at the wavelength of the peak sensitivity.

$$D^* (D\text{-star}) = \frac{[Effective\ Sensitive\ Area\ (cm^2)]^{1/2}}{NEP}$$

Short Circuit Current (Ish)
This value is measured using white light of 2856K distribution temperature from a standard tungsten lamp of 100 lux illuminance. The short circuit current is that current which flows when the load resistance is 0 and is proportional to the device photosensitive area.

Dark Current (Id) and Shunt Resistance (Rsh)
The dark current is the small current which flows when reverse voltage is applied to a photocell under dark conditions. It is a source of noise for applications in which a reverse bias is applied to photocells, as is the case with PIN photocells. To observe the dark current there are two methods—observation of the V/I ratio (termed shunt resistance) in the 0V region (−10mV for the data herein), or observation of the current at actual applied reverse bias conditions.

$$R_{sh} = \frac{10\ (mV)}{Dark\ Current\ at\ V_R = 10mV\ (A)}$$

Junction Capacitance (Cj)
An effective capacitor is formed at the P-N junction of a photocell. Its capacitance is termed the junction capacitance and is the major factor in determining the response speed of the photocell. This is measured at 1MHz for PIN types and 10kHz for other types.

Rise Time (Tr)
This is the measure of the photocell's response to a stepped incident light input. It is the time required to transition from 10% to 90% of the normal high level output value. Since the rise time is a function of the wavelength of the incident light and of the load resistance, for the purposes of the specifications given herein, the value for the case of either a GaAsP (655nm) or GaP (560nm) LED source and a 1kΩ load are used, except for the case of PIN photocells.

Frequency Response
This is the measure of the photocell's response to sine-wave incident lights and frequently used for PIN photocells. It is defined as the frequency at which the output current decreases by 3dB of the low frequency response. The load resistance used is 50Ω.

Maximum Reverse Voltage (VR max)
Applying reverse voltages to photocells can cause breakdown and severe deterioration of device performance. Therefore reverse voltage should be kept somewhat lower than the maximum rated value, VR max, even for instantaneously applied reverse bias voltages.

Construction and Operating Characteristics

INTRODUCTION

Silicon and GaAsP photocells make use of the photovoltaic effect—the generation of a voltage across a P-N junction when the junction is exposed to light. Photocells have been used in such devices as computer card readers and smoke detection systems and are now finding use in exposure meters, analysis equipment and radiation measurement equipment by virtue of recent improvements in S/N ratio, response speed and linearity. Some major features of photocells include:

1) **High speed response**
2) **Excellent linearity**
3) **Small dark current (low noise)**
4) **Wide spectral response**
5) **Mechanical ruggedness**

This section will serve to introduce the major aspects of photocell construction and operating characteristics.

CONSTRUCTION

Hamamatsu photocells listed in this catalog can be classified by construction into 4 types of silicon photocells and 2 types of GaAsP photocells.

TABLE 1
Photocell Construction

Type	Construction	Photocell Types
Planar Diffusion Type		S874, S875 Series S1087, S1133 Series Diffusion Type GaAsP
Low C_j Planar Diffusion Type		S1336 Series S1337 Series
PNN$^+$ Type		S1226 Series S1227 Series
PIN Type		PIN Silicon Photocells
Schottky Type		Schottky Type GaAsP

* Prefix letter S identifies silicon and G GaAsP.

● Planar Diffusion Type

An SiO_2 coating is applied to the P-N junction surface, yielding a photocell of low level dark current when compared with conventional diffusion type photocells.

● Low Capacitance Planar Diffusion Type

A high speed version of the planar diffusion type device, this type makes use of a highly pure, high resistance N-type material to enlarge the depletion layer and thereby decrease the junction capacitance, thus lowering the response time to 1/10 the normal value. The P-layer is made extra thin for high ultraviolet response.

● PNN$^+$ Type

A low resistance N$^+$ material layer is made thick to bring the N-N$^+$ boundary close to the depletion layer. This somewhat lowers the sensitivity to infrared radiation, making this type of device useful for measurements of short wavelengths.

● PIN Type

An improved version of the low capacitance planar diffusion type device, this type makes use of an extra high resistance I-layer between the P- and N-layers to improve response time. This type of device exhibits even further improved response time when used with reverse bias and so is designed with high resistance to breakdown and low leakage for such applications.

● Schottky Type

A thin gold coating is sputtered onto the N material layer to form a Schottky Effect P-N junction. Since the distance from the outer surface to the junction is very small, ultraviolet sensitivity is high as in the case of other UV enhanced photocells.

THEORY OF OPERATION

Figure 1 (a) shows a cross section of a photocell with the dark condition band model given at (b). Since this device is in thermal equilibrium, the P-layer and N-layer Fermi Levels are equal and a voltage gradient develops in the depletion layer by virtue of the contact potential (potential barrier).

FIGURE 1
Photocell in the Dark Condition

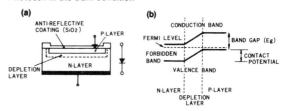

When radiation is allowed to strike the photocell, the internal electrons become stimulated. If the stimulating radiation is of high enough level, i.e., larger than the band gap, Eg, electrons will be pulled into the conduction band, leaving behind positive holes. These electron/hole pairs are generated throughout the N and P material and the depletion layer. In the depletion layer, electrons and holes drift towards the N and P material layers respectively. In addition, electrons in the P-layer diffusion length and holes in the N-layer diffusion length diffuse towards the depletion layer, causing a charge to accumulate in the N- and P-layers (Figure 2 (a) and (b)). This is known as the photovoltaic effect. Sensitivity can be improved by making the P-layer as thin and transparent as possible and making the depletion layer as wide as possible.

As a natural reaction to the lower P-N potential barrier, electrons and holes drift towards the P and N materials respectively, reducing the accumulated charge. For this reason continued application of radiation will not result in unlimited buildup of charge—the potential barrier is reduced until apparent charge transfer ceases. This potential barrier change, termed Fermi Level Shift, manifests itself externally as a voltage at the photocell terminals. This voltage is the open circuit voltage, Vop in (c). If the terminals of the photocell are shorted, the current which flows is known as the short circuit current Ish in (d).

FIGURE 2
Photocell with Incident Radiation Present

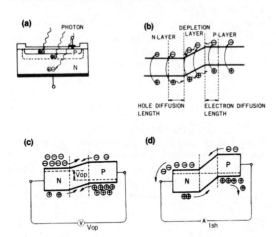

V-I CHARACTERISTICS
When a voltage is applied to a photocell in the dark state, the V-I characteristic curve observed is similar to the curve of a conventional diode as shown in Figure 3 (a).

FIGURE 3
V-I Characteristics

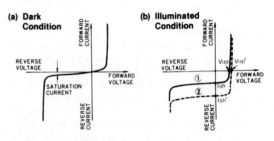

When light strikes the photocell, however, the curve shifts to position ① in Figure 3 (b). Increasing the amount of incident light shifts the characteristic curve still further to position ② of that figure. Under such conditions an open circuit will exhibit a voltage Vop (or Vop') while shorting the photocell terminals will result in the flow of the short circuit current Ish (or Ish'). The open circuit voltage is a forward voltage in the sense used with conventional diode rectifiers. Figure 4 shows the relationship between incident light and this forward voltage and short circuit current.

FIGURE 4
Incident Light vs. Output Signal Relationship (S874-5K)

(a) Open Circuit Voltage (Vop)

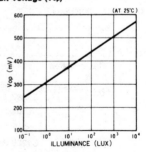

(b) Short Circuit Current (Ish)

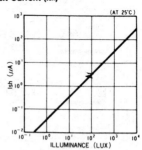

Note that the Ish characteristic is quite linear with the range of linearity achievable reaching 6 to 8 orders of magnitude depending on the type of device. For this reason precision measurements of light using photocell rely not on the open circuit voltage but rather on the current produced. Use a negatively feedbacked operational amplifier (OP-amp) with a near zero input impedance. While, the V_{op} versus incident light relationship is logarithmic but is greatly affected by variations in temperature making it unsuitable for light intensity measurement applications. The usual method of achieving logarithm compression is to use a logarithmic diode as the feedback element in an OP-amp circuit (see Figure 5).

FIGURE 5
OP-Amp Connection Examples

(a) V-I Conversion

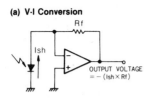

(b) Log Response Amplifier

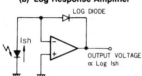

When connecting with a finite load resistance R, care should be taken with the design since linearity is affected by R. Smaller values of R result in operation close to the Ish curve and a wide linear operating range (see Figure 6).

FIGURE 6
Load Lines

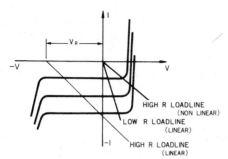

HIGH R LOADLINE (NON LINEAR)
LOW R LOADLINE (LINEAR)
HIGH R LOADLINE (LINEAR)

Equivalent Circuit

A photocell equivalent circuit is shown in Figure 7.

FIGURE 7
Photocell Equivalent Circuit

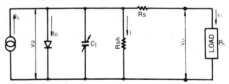

IL : current generated by the incident light
ID : diode current
Cj : junction capacitance
Rsh : shunt resistance
Rs : series resistance
I' : shunt resistance current
VD : voltage at the diode terminals
Io : output current
Vo : output voltage

Using the above equivalent circuit for analysis and solving for the output current, we have:

$$I_o = I_L - I_D - I' = I_L - I_s \left[\exp \frac{qV_D}{kT} - 1 \right] - I'$$

Where Is : photocell reverse saturation current
q : electron charge
k : Boltzmann's constant
T : absolute temperature of the diode

The open circuit voltage V_{op} is the output voltage when Io = 0. Therefore we have:

$$V_{op} = \frac{kT}{q} \ln \left[\frac{I_L - I'}{I_s} + 1 \right]$$

If we ignore I', V_{op} is logarithmically proportional to IL and Is increases exponentially with increasing ambient temperature, making V_{op} inversely proportional to the ambient temperature. This relationship does not, however, hold for very small amounts of incident light such as "IL ≈ Is" region.

Ish is the output current when the load resistance RL = 0 and Vo = 0, yielding:

$$I_{sh} = I_L - I_s \left[\exp \frac{q(I_{sh} \cdot R_s)}{kT} - 1 \right] - \frac{I_{sh} \cdot R_s}{R_{sh}}$$

where the 2nd and 3rd terms above limit Ish linearity. If Rs is several ohms or less and Rsh is 10^7 to 10^{11}, these terms become negligible over quite a wide range.

SPECTRAL RESPONSE CHARACTERISTICS

As explained in the section on principles of operation, when the energy of absorbed photons is lower than the band gap Eg, the photovoltaic effect does not occur. For silicon photocell the band gap energy is 1.12eV and for GaAsP it is 1.8eV resulting in a long wavelength cutoff of 1100 and 700nm respectively for the two types of photocells. Short wavelength sensitivity is, however, determined by the P-layer thickness, increasing as the layer's thickness decreases. While normally this cutoff is 300 to 400nm, photocells intended for use in the UV region can be produced with extra thin P-layers to obtain cutoffs below 200nm.

Figure 8 shows the spectral response curves for S1336 and S1337 series silicon photocells. BQ types have a fused silica window, BK types a borosilicate glass (Kovar glass) window and BR types resin coated window. Other photocells' spectral response curves are shown along with tabulated data.

FIGURE 8
Spectral Response Characteristics of S1336 and S1337 Series

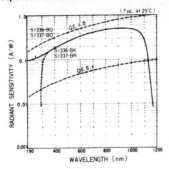

FIGURE 9
Photocell Biasing Circuit

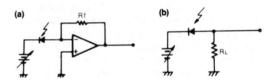

FIGURE 10
Rise Time vs. Bias Voltage

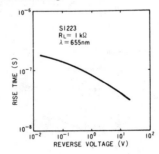

FIGURE 11
Linearity Limits

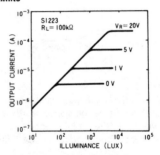

APPLICATION OF REVERSE BIAS

Although photocells generate a photovoltaic voltage output without the application of an external power source, speed of response and linearity can be improved by the use of such an external biasing source. It should be borne in mind that the signal current flowing in a photocell circuit is determined by the number of photovoltaically generated electron/hole pairs. Thus the application of a bias voltage does not result in the loss of photoelectric conversion linearity.

Figure 9 shows examples of circuits in which a photocell is reverse biased. The photocell acts not as a voltage source but rather as a photoconductive circuit element. Figure 10 and 11 show the effect of bias voltage on response speed and linearity limits respectively.

While application of bias to a photocell is very useful in improving response speed and linearity, it has the accompanying disadvantage of increasing dark current and noise levels along with the danger of damage to the photocell caused by excessive bias voltage. PIN devices designed for such high speed applications have excellent dark current as well as a high resistance to reverse voltages.

NOISE CHARACTERISTICS

As in the case with other types of light sensitive devices, the limits of light detection sensitivity are determined by the noise characteristics of the photocell. The two major sources of noise are Johnson noise of the shunt resistance and dark current shot noise.

When used with zero bias the Johnson noise, i_j, only is present, as given by;

$$\overline{i_j^2} = 4kTB/R_{sh}$$

> where k : Boltzmann's constant
> T : absolute temperature
> B : noise bandwidth

When a bias is applied, however, shot noise, i_s, becomes the major noise component, according to the relationship;

$$\overline{i_s^2} = 2qIB$$

> where q : electron charge
> I : signal current + dark current
> B : noise bandwidth

Both noise components exhibit a flat spectrum and are expressed in units of $A/Hz^{1/2}$. While it is usual to express the lower limit of detection sensitivity as the equivalent amount of light required to produce a signal current equal to the noise current, sensitivity itself varies with wavelength so that the wavelength must be specified when giving this light level. For the purposes of this catalog, the wavelength of peak sensitivity is used for the NEP (Noise Equivalent Power) specifications with zero bias.

$$NEP = i_j/S \quad (W/\sqrt{Hz})$$

> where i_j: Johnson noise
> S: peak sensitivity

With DC type measurement techniques, photocell and amplifier drift and low frequency flicker cause variations in the measured value such that the overall effective NEP is a few orders of magnitude increased over the device NEP specification. By using a chopper to mechanically modulate the light signal and synchronously detect only the chopping frequency, low frequency noise components can be eliminated, yielding noise performance close to the device NEP.

FIGURE 12
Synchronous Measurement

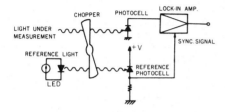

RESPONSE SPEED

Response speed is the measure of the time required for the P- and N-layer accumulated charge to become an external current. It is basically determined by the junction capacitance C_j, series resistance R_s, and the load resistance R_L. It is common to express the response speed as the rise time, t_r, from 10% to 90% of the normal output value upon applying a step input light signal. t_r is given by;

$$t_r = 2.2C_j(R_L + R_s)$$

where, for $R_L \gg R_s$, $t_r \approx 2.2C_jR_L$. While making C_j and R_L small will improve speed of response, R_L is usually determined by external factors and cannot be made arbitrarily small for this purpose. C_j is directly proportional to the photosensitive area A and inversely proportional to the square root of the bias voltage V_R and the resistivity of the substrate material ρ.

$$C_j \propto A/\sqrt{V_R}\rho$$

Therefore achieving fast response time is possible by applying a bias, making the photosensitive area small and using a high resistivity substrate material. PIN silicon photocells use such a substrate material to achieve high speed response.

FIGURE 13
Rise Time vs. Load Resistance with Photosensitive Area as Parameter

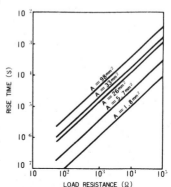

TEMPERATURE CHARACTERISTICS

Ambient temperature variations greatly affect photocell sensitivity and dark current. The cause of this is variation in the light absorption coefficient which is temperature related. For long wavelengths, sensitivity increases with increasing temperature (positive temperature coefficient) and for short wavelengths it decreases (negative temperature coefficient). Photocells for use in the UV region have low absorption for the shorter wavelengths and thus less temperature dependence in this region. Figure 14 shows the temperature coefficient for planar diffusion photocells and low capacitance UV enhanced type photocells (S1336, S1337 series).

FIGURE 14
Temperature Coefficient vs. Wavelength

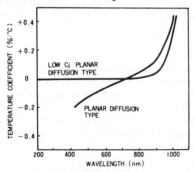

In addition to the above effect, increasing temperature causes electrons in the valence band to become excited and pulled into the conduction band thus resulting in a constant increase in dark current with increasing temperature. As shown in Figure 15 (a), an increase of 10°C causes an increase in dark current of 2 to 4 times. This is equivalent to a similar reduction of Rsh and the subsequent increase in Johnson and shot noise (see section on noise). (b) shows an example of Vop temperature dependence.

FIGURE 15
(a) Dark Current Temperature Dependence

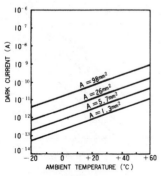

(b) Vop Temperature Dependence

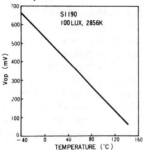

USE OF OPERATIONAL AMPLIFIERS

Since the equivalent input resistance of an OP-amp circuit (see Figure 16), being the feedback resistance divided by the open loop gain, is very low the Ish mode of operation can be used to achieve a high degree of linearity. Let us examine some considerations for the use of OP-amps with photocells.

● Feedback Circuit

Figure 16 shows a typical OP-amp circuit. The input current, Ish, is converted to an output voltage equal to Ish × Rf. To eliminate the effect of noise, the feedback resistance Rf should be made as small as possible with respect to the photocell shunt resistance Rsh. Cf serves as a damping capacitance to prevent oscillation, with several pF being sufficient for this purpose. The feedback network also acts as a noise filter with a time constant Rf × Cf and tends to limit the response speed of the circuit, making selection of component values for this network an important design consideration.

For use with very small amounts of light, Rf is made large, sometimes as large or larger than Rsh. It should be borne in mind that under these circumstances the noise, multiplied by (1 + Rf/Rsh) appears in the circuit's output along with the signal voltage output.

Offset errors can be reduced to below 1mV by use of a variable resistor as an external offset adjustment.

FIGURE 16
Basic OP-Amp Circuit

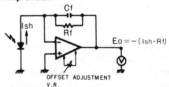

● Bias Current

Although the OP-amp input resistance is much greater than the value of Rf and all the input current Ish might be thought of as flowing through the resistance Rf, in reality the finite input impedance of the OP-amp results in some bias current flowing and subsequent measurement errors. For bipolar OP-amps this bias current, being several nA to several hundred nA, is large compared to the current Ish, making an FET input OP-amp the ideal choice for use in very low light level applications, FET input OP-amps have bias currents in the region of 10pA or less. Bias current, like offset, can be adjusted out with an external variable resistor if one is provided in the circuit design.

● Gain Peaking

Although the bandwidth of a photocell OP-amp circuit is determined largely by the Rf × Cf time constant, gain peaking sometimes occurs. Such peaking, as shown in Figure 17, is caused by phase differences between Ish and

E_0. While the occurance of gain peaking could be predicted by solving the gain transfer function, this calculation is quite complex. A rough idea of gain peaking effects can be had by using graphical methods (see Figure 18). For an actual test of whether gain peaking is present, apply a rectangular light input signal to the photocell and observe the output waveform. Gain peaking is indicated by a ringing on the output waveform.

FIGURE 17
Gain Peaking

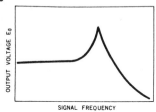

Graphical Gain Peaking Analysis
FIGURE 18

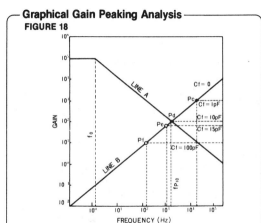

Figure 18 is an example of graphical gain peaking analysis. R_{sh} and C_j are taken to be 1GΩ and 1nF respectively and R_f is taken as 1MΩ. The value of C_f is varied and gain peaking is checked for graphically. Line A represents the OP-amp open loop gain (DC gain $A_0 = 10^5$, $f_0 = 1.6$Hz) and line B the closed loop gain with $C_f = 0$. When C_f is given finite values of capacitance the closed loop gain to the right of the resultant poles (frequency $f_p = 1/C_f \cdot R_f \cdot 2\pi$) is constant with frequency increase. The example shows the graphs for $C_f = 1$, 10, 15 and 100pF. The relative positions of the closed loop gain poles and the line A indicates whether or not gain peaking will occur. Whenever the gain at the pole exceeds the open loop gain, peaking occurs at the frequency of the intersection of line A with line B. In the example shown, gain peaking occurs for values of $C_f \leq 10$pF at a peaking frequency of 1.6 (kHz).

RELIABILITY AND LIFE

Hamamatsu photocells, if used within the operating conditions described in this catalog should rarely exhibit any deterioration of performance. Most cases of such deterioration are results of package, lead or filter failures. Table 2 gives a summary of the Hamamatsu photocell reliability and life testing standards.

TABLE 2
Test Standards

Category	Conditions	EIAJ Standards	MIL Standards
Soldering Resistance	260°C for 10 sec.	SD-121, A-1	202D-210B
Solderability*	230°C for 5 ~ 10 sec.	A-2	202D-208B
Thermal Shock	100°C for 15 sec./ 0°C for 5 sec.: 5 cycles	A-3	750B-1056. 1
Thermal Cycling	−20°C for 30 min./ 60°C for 30 min.: 5 cycles	A-4	
Shock	100G for 6 msec. X, Y, Z directions	A-7	
Drop	Three drops onto wooden surface from 75 cm	A-8	
Vibration	100~2000Hz, 20G	A-10	750B-2056
Terminal Strength	Pulling 0.5 kg Bending two times Twisting two times	A-11	
High Temp. Storage*	80°C, 100°C	SD-121, B-9	
High Temp/ Humidity Life Test*	60°C, 80°C, 95%RH	B-10	
Low Temp. Storage	−55°C	B-11	
High Temp. Operating*	80°C with bias applied		

*Conditions may be limited by photocell package.

FIGURE 19
High Temperature/Humidity Storage Testing

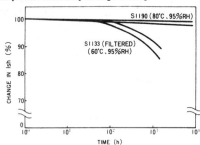

PHOTOCELL SELECTION

● Wavelength

In the usable wavelength range of 190 to 1200nm, spectral response varies different types of photocell as shown in Selection Guide. It is important to select a photocell the response of which matches the light to be detected.

TABLE 3
Photocell Selection on Wavelength

Wavelength Range	Photocell Type
UV, Blue (190~400 nm)	S1226, S1227 S1336, S1337 S1722-02, S1723-02 Schottky Type GaAsP
Visible (400~700 nm)	S1087, S1087-03 S1133, S1133-03, S1133-05 Diffusion Type GaAsP
Red, IR (longer than 650 nm)	Almost all silicon photocells
Low IR sensitivity	S1226, S1227 S1087, S1087-03 S1133, S1133-03, S1133-05 GaAsP Photocells

● Light Energy

Since photocell noise is proportional to the photosensitive area, the limits of detection sensitivity are best reached by making this area as small as possible for low noise operation. For applications requiring the detection of very low level light, this should be taken into consideration along with some form of light gathering lens or mirror system.

In general, DC type measurement techniques will result in an effective overall NEP of a few orders of magnitude worse than that listed in this catalog. Although the effect of the load resistance can cause saturation for large amounts of light, photocells can be used with excellent linearity up to several tens of thousand lux light levels.

● Response Speed

For reception of fast changing light signals, the photocell response speed must be considered. Response speed ranges from sub- μs to several tens of μs and is determined by the junction capacitance and load resistance, smaller values of each improving the response speed. PIN devices are designed with an especially small junction capacitance and have good response speed. In addition the S1336 and S1337 devices have extra low junction capacitance and response times in the region of 0.2 to 3 μs.

The required rise time to detect a light signal changing with a frequency of f is given by:

$$tr \leq \frac{0.35}{f}$$

where f is in herts and tr in seconds.

● Environmental Conditions

After being subjected to high temperature or humidity photocells often exhibit a drop in the value of R_{sh} due to leaks in the package. The best packages are metallic cases while ceramic case devices have slightly more limited temperature and humidity ranges. For filtered photocells, the filter itself determines the allowable enviromental conditions for the device (see Figure 19).

● Incident Light Beam

The photocell device to be used is also determined somewhat by the size of the light beam to be detected. The determining factor is the photosensitive area which, depending on the device type, ranges from 1.2mm² to 100mm². A device with a photosensitive area larger than necessary should be avoided since the excess area will result in increased noise and in general, deteriorated performance. In such cases, some form of light gathering optical system is desirable to allow small area devices to be used.

For alarm applications devices with lens type windows are the ideal choice because of their directional characteristics.

Precautions for Use

● Window

Care should be taken not to touch the window with the bare hands, especially in the case of UV photocells since foreign materials on the window can seriously affect transmittance in the UV range.

Ethyl alcohol should be used to clean the light window. Other type cleansing agents could cause deterioration of the device's resin coating or filter and should be avoided.

Lightly wipe dirt off the window using ethyl alcohol.

● Soldering

Since photocells are susceptible to damage from excess heat, care must be given to temperature and dwell time when soldering such devices. As a rule, metal case devices should be soldered at 260°C or below within 10 seconds and ceramic case devices at 260°C or below within 5 seconds. For soldering small devices, some form of heat sinking such as the use of a pair of tweezers to hold the device leads while soldering is also recommended.

Mount ceramic case types 5 mm minimum away from any surface and solder at 260°C maximum for 5 seconds maximum time.

Use tweezers, etc. as a heatsink when soldering small photocell.

● **Lead Handling**

Care should be taken to keep within the recommended mechanical limits: 0.5kg pulling strain, two 90°C bends and two twists of the device leads at 6mm minimum away from the body.

Warranty

As a general rule Hamamatsu photocells are warranted

for one year after delivery. The warranty is limited to replacement of the faulty device. It does not cover cases of operational failure caused by accident or misuse of the devices.

Replacement Type Devices

The SPC series (visible to near infrared: type No. S639, S640, S641, S642, S643, S740, S827, S851), SPC-B series (UV to near infrared: type No. S639B, S640B, S641B, S642B, S643B), S780 series and S876 series while not listed in this catalog are not discountinued devices. They are available upon request and since they are not normally stocked, delivery will take somewhat longer than other standard device types. New designs should use the devices listed in this catalog to assure best availability.

CIRCUIT THEORY

The reversed biased (PIN) photodetector produces a voltage across bias resistor (R1) and capacitively couples it to the gate of field-effect transistor (Q1) used as a signal preamplifier. See Fig. 2-1. The response of the PIN photodetector is determined by the resistance of R1 and is relatively low due to the nanosecond rise time of the laser pulses detected. The output of Q1 is further amplified by a wideband amplifier (A1). The bandwidth of this stage is over 70 MHz with a gain of up to 200 being controlled by trimpot (R16).

The output of A1 is fed to jack (J1) and will reproduce these narrow detected pulses. They can now be used for further amplification or processing by other external circuitry. The output of A1 is also further amplified by transistors (Q2 and Q3) for triggering the timer (A2). This stage expands these narrow detected pulse widths to where they can be used for listening (via headsets) or to drive external circuitry such as relay drivers, counters, etc. A threshold adjustment for this stage is set by trimpot R7. The now expanded width or duty cycle is controlled by trimpot R12. The circuit is built on an assembly board as shown in Fig. 2-2 and is intended to be positioned into a cylinderical housing with a lens and IR filter. Power for the device is via three preferably rechargeable Nicad batteries (B1, 2, 3) being controlled by switch (S1A, B). Note that B3 only supplies reverse bias current to the PIN and need not be turned "off" due to the low current drain. Note that conventional batteries may be used but they should have D3 and D4 to reduce the voltage. Remember that four nicad cells equals 4.8 volts and four conventional cells equals 6.0 volts.

CONSTRUCTION STEPS

1. Layout, identify and match all components with the Table 2-1 parts list.
2. Position components as shown in Fig. 2-2 and observe proper polarity of diodes, semiconductors and electrolytic capacitors. Wire as shown using leads of com-

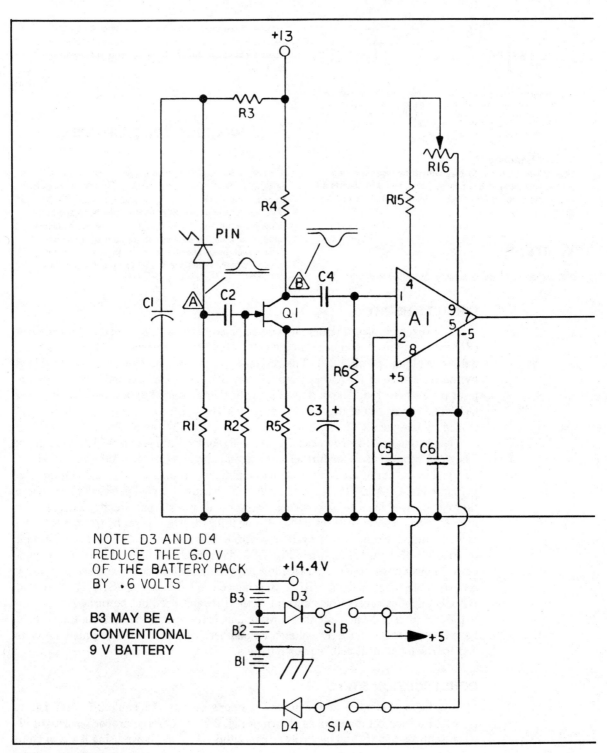

Fig. 2-1. Schematic diagram.

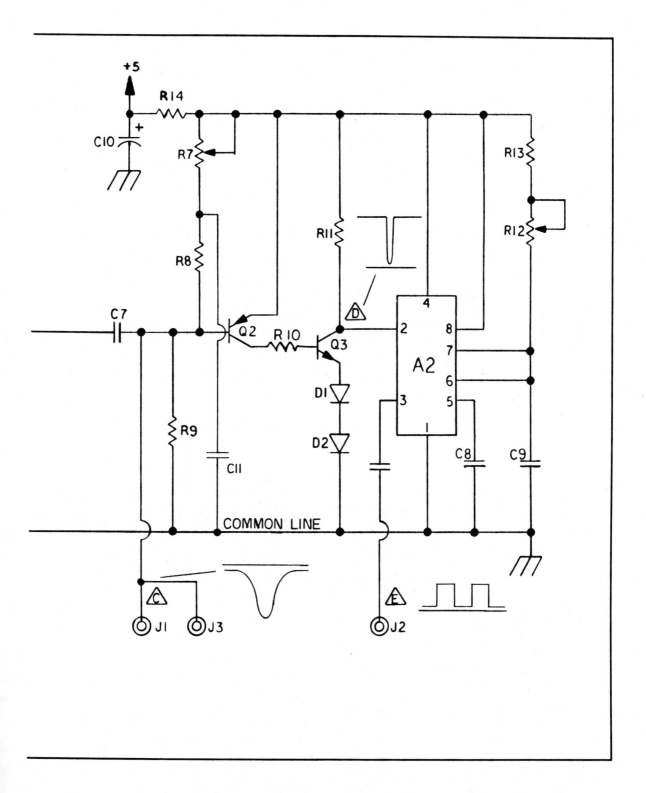

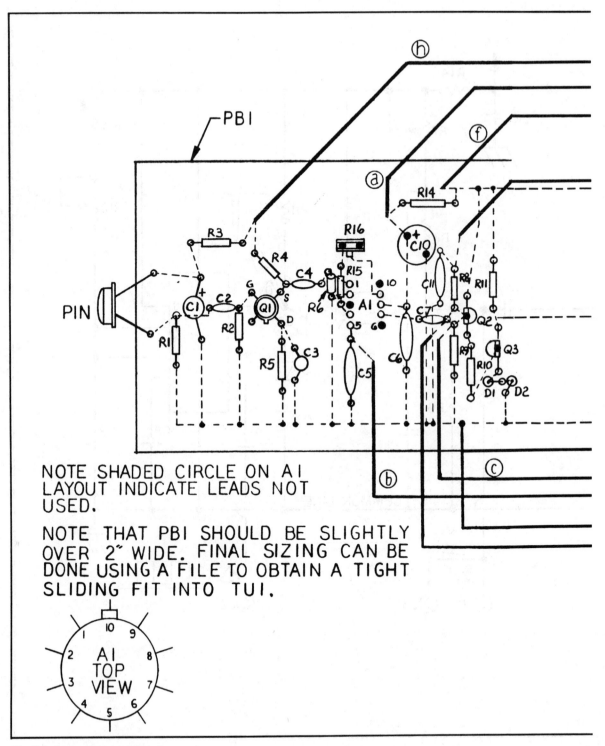

NOTE SHADED CIRCLE ON A1
LAYOUT INDICATE LEADS NOT
USED.

NOTE THAT PB1 SHOULD BE SLIGHTLY
OVER 2″ WIDE. FINAL SIZING CAN BE
DONE USING A FILE TO OBTAIN A TIGHT
SLIDING FIT INTO TU1.

Fig. 2-2. Assembly board layout.

32

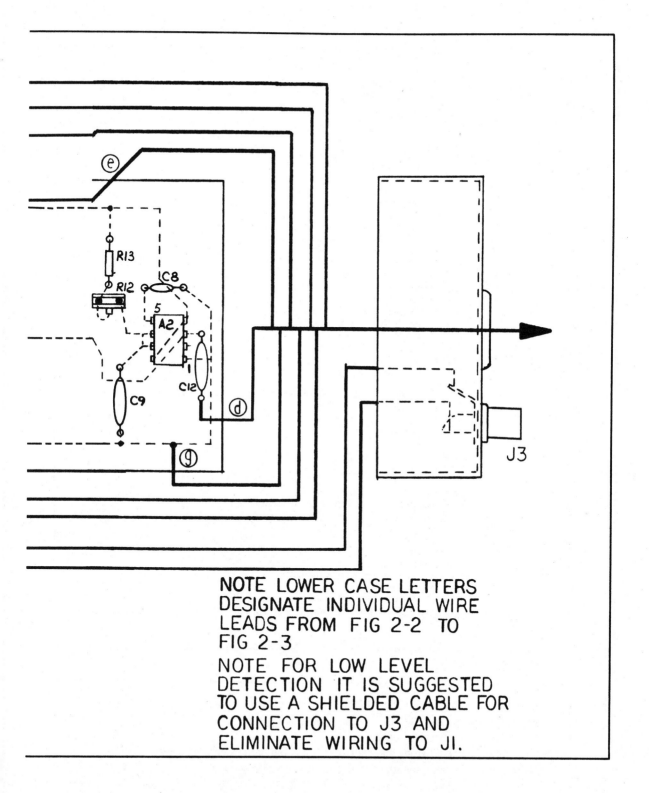

NOTE LOWER CASE LETTERS
DESIGNATE INDIVIDUAL WIRE
LEADS FROM FIG 2-2 TO
FIG 2-3

NOTE FOR LOW LEVEL
DETECTION IT IS SUGGESTED
TO USE A SHIELDED CABLE FOR
CONNECTION TO J3 AND
ELIMINATE WIRING TO JI.

Table 2-1. LSD3 Laser Detector Parts List.

Part	Qty	Description
R1,3 8,13	4	1-kΩ 1/4-watt resistor
R2	1	470-kΩ 1/4-watt resistor
R4,5,10 11	4	4.7-kΩ 1/4-watt resistor
R6	1	10-kΩ 1/4-watt resistor
R12,16	2	2-kΩ trimpot
R7	1	100-kΩ pot for threshold adjust
R9	1	1-MΩ 1/4-watt resistor
R14	1	10-Ω 1/4-watt resistor
R15	1	100-Ω 1/4-watt resistor
C1	1	10-μF/25-V electrolytic cap
C2,4,7	3	.001-μF/25-V disc cap
C3	1	.47-μF tantalum or electrolytic
C5,6,9 11,12	5	1-μF/25-V disc cap
C8	1	.01-μF/25-V disc cap
C10	1	100-μF/25-V electrolytic cap
D1,D2	2	1N914 silicon signal diodes
D3,4	2	1N4001 50-V rectifier see text Fig. 2-1
Q1	1	2N4416 FET general-purpose transistor
Q2	1	PN2907 PNP GP transistor
Q3	1	PN2222 NPN GP transistor
A1	1	Wideband op amp LM733
A2	1	Timer 555
S1A,B	2	DPST slider switch
CL1,2,3	3	Battery clips
PB1	1	4-1/4" × 2" perfboard adjust width for tight sliding fit into TU1
J1,2,3	3	RCA phono jacks
HES1	1	Heatsink for A1 (may not be necessary if used at 5 V)
PIN	1	Photodiode detector (see text)
EN1	1	Small metal box 5 × 3 × 2
TU1	1	2-3/8" × 8 " PVC tubing
FIL	1	IR filter
LE1	1	54 × 89 lens
CA1,2,3	3	2-3/8" plastic cap #A 2-3/8" (see text)
TAP1		Small pieces of 2 sided foam tape or use adhesive
WR1	6'	#24 hook-up wire vinyl
BU1,2	2	Small 1/2" bushing or grommet
SW1/NU1	2	6-32 × 1" screws/nuts
BH1,2	2	4 AA cell holders
WN1	2	Small wire nuts
HS1	1	Optional headsets with match transformers
B1,2	8	Use 4 AA cell each
B3	1	7.2-to 9-V transistor battery
B1,2,3	3	(4) 1.2-V nicads (see note on Fig. 2-1) or use (4) 1.5-V cells for each B1 and B2

Complete kit of above or assembled and tested unit is available from Information Unlimited, P.O. Box 716, Amherst, NH 03031. Write or call (603) 673-4730.

ponents for connections. Avoid wire bridges. Note that layout can be more compact than that shown, but it should be followed as nanosecond pulses are present and all attempts to limit stray inductance is advised.

3. Attach wires and connect to external controls, jacks, and battery clips as shown in Fig. 2-3. Note lowercase letters for identifying interconnecting wires.

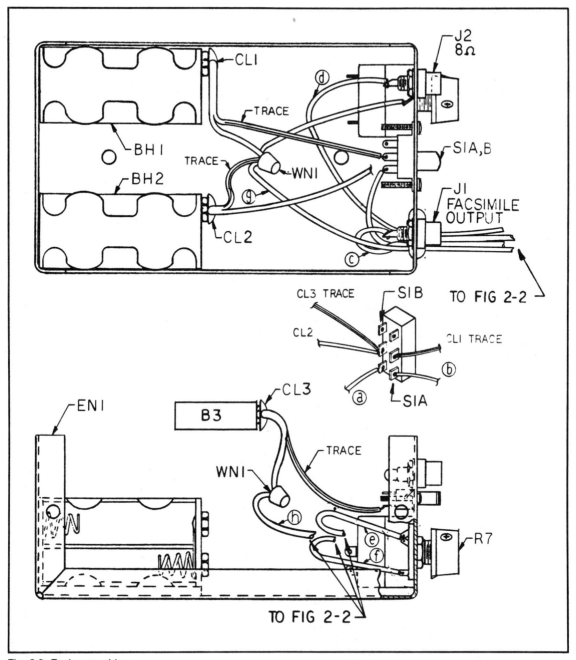

Fig. 2-3. Enclosure wiring.

    4. Fabricate power and control enclosure as shown in Figs. 2-4 and 2-5. Follow the approximate layout shown. Note front panel controls.

    5. Fabricate the main housing from a piece of PVC as shown in Fig. 2-4.

    6. Fabricate cap (CA1) for bushing (BU2) and jack (J3).

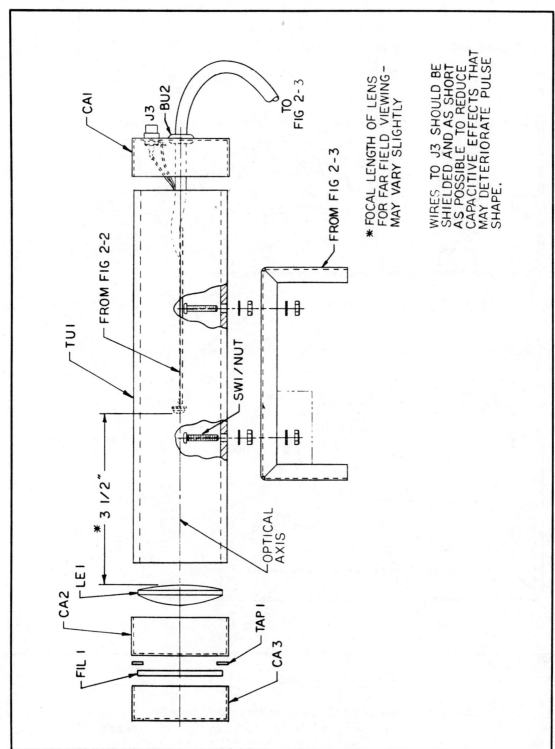

CA1

J3
BU2

TO
FIG 2-3

FROM FIG 2-3

* FOCAL LENGTH OF LENS
FOR FAR FIELD VIEWING-
MAY VARY SLIGHTLY

WIRES TO J3 SHOULD BE
SHIELDED AND AS SHORT
AS POSSIBLE TO REDUCE
CAPACITIVE EFFECTS THAT
MAY DETERIORATE PULSE
SHAPE.

FROM FIG 2-2

TU1

SW1/NUT

OPTICAL
AXIS

* 3 1/2"

LE1

CA2

FIL 1

TAP 1

CA 3

Fig. 2-4. Main housing and optics.

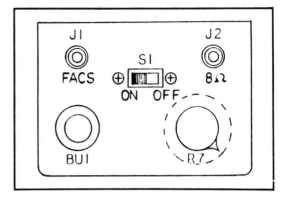

Fig. 2-5. Front panel.

7. Fabricate caps (CA2, CA3) by removing center. This is easily done by sleeving them onto TU1 and cut away center using the ID of the tube as a guide. Note CA2 retains lens (LE1) to TU1 while CA3 retains the infrared filter (FIL1). Small pieces of two sided tape (TAP1) also may be used.

8. Final assembly is as shown in Figs. 2-4 and 2-6. Note screws (SW1/NU1) shown for securing housings together. Note that perfboard (PB1) should sleeve into TU1 tightly. This automatically will position the PIN at the center of the optical axis and allow for adjustment of the distance to the lens.

9. Testing of the unit should be performed with the assembly board out of the enclosure tubing TU1. Check all wire and solder for accuracy and quality of joints.

10. Obtain a laser pulse to transmitter similar to our model SSL3/5 and place down-range about 30 to 40 feet. Remove all external optics and point the laser emitter in the general direction of the area where the unit is to be tested. Set at a 1k to 2k pulse rate.

11. Position and secure the assembly board with the PIN diode facing the laser transmitter.

12. Install the batteries and check voltages on the assembly board as shown in Fig. 2-1.

13. Connect scope to point (A) Fig. 2-1 and note a pulse as shown approximately 200 nanoseconds wide and at an amplitude of 5-10 millivolts. (Note pulse should be a good reproduction of that of laser pulses.)

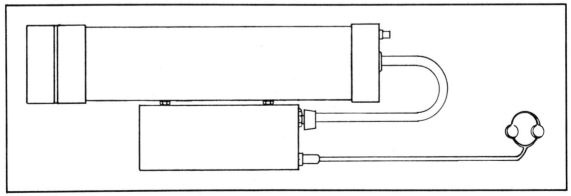

Fig. 2-6. Final assembly.

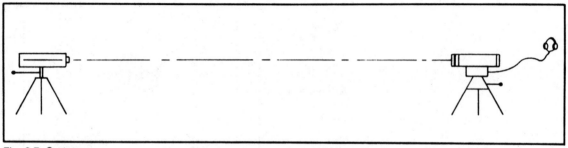

Fig. 2-7. System.

14. Measure point (B) and note same shaped inverted pulse of 50-100 millivolts.

15. Measure point (C) and note a 0.5 to 1 volt pulse. Note adjustment of R16 controlling the gain of amplifier A1.

16. Measure point D and note a negative pulse as shown of four volts and 10 to 20 microseconds wide.

17. Measure point E and note a 4-volt waveshape as shown. Note adjustment of R7 controlling the threshold of triggering for A2. R12 adjust the pulse symmetry.

It should be noted that the output at J2 is for monitoring the output via headsets or for powering a relay or other controlled devices. The actual pulse shape at J1 and J3 is the nanoseconds and is a reproduction of the original laser pulse for further processing etc.

18. Reassemble the unit and "range check" using setup shown in Fig. 2-7. A maximum range of many kilometers is possible between the transmitter and receiver. The sensitivity is greatly increased by the optics. Further sensitivity is possible using a larger lens in the receiver. A range of 12 kilometers was obtained by the original builder with power to spare. A tripod was necessary for both transmitter and receiver with optical alignment being relatively critical for this range.

A SPECIAL NOTE ABOUT PIN DIODES

It is suggested to use one of two PIN devices for this circuit. A low cost approach utilizes the Hamamatsu #S1223-02 that provides reasonable performance. Those wishing to obtain optimum performance may wish to use the more expensive Hamamatsu SS1336BK. This device is more sensitive when detecting high speed laser diode pulses.

Chapter 3

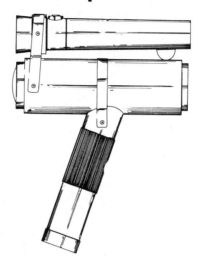

See In The Dark
IR Viewing Device (SD5)

T HIS PROJECT SHOWS HOW TO CONSTRUCT A DEVICE CAPABLE OF ALLOWING ONE
to see in total darkness. It can be used to view a subject for recognition or evi-
dence gathering reasons without any indication that he is under surveillance. Many
uses such as detection, alignment of IR alarms, laser communications systems and other
IR sources make it an invaluable device in these fields. Other uses include detecting
diseased vegetation in certain types of crops from the air, an aid to night time varmit
hunting as well as viewing high temperature thermographic scenes providing the color
temperatures are sufficient. This device is excellent for use with our SSL3 laser described in Chapter 1 allowing complete viewing of this optical energy.

The unit is built with cost and performance as the objective. It is as good opera-
tionally as units costing three to five times as much, yet is lighter and more versatile
than the more expensive ones. The batteries are enclosed in the housing and do not
require side packs, cables etc. The range and field of viewing are determined by the
intensity of the intergrated IR source and viewing angle of the optics. While the optics
offered in the basic kit are usable they are not the best and will have spherical aberra-
tion and other adverse effects. This approach keeps the basic cost down for those not
requiring actual viewing of detailed scenes. Improved optics will eliminate these effects and can be obtained as an option.

Assembly is around commonly available PVC tubing as the main housing and a
specially designed patented miniature power source for energizing the image tube. The
type of tube is a readily available image converter similar to that being used by most
manufacturers of similar devices. This tube establishes the limits of viewing resolution and is suitable for most applications but may be limited if one desires video per-

fection. This usually applies to similar devices and is the limit of quality of performance available assuming standard optics are used.

The viewing range is determined mainly by the intensity of the IR source and can be controlled by varying this parameter. Our basic unit is shown utilizing a 2 D-cell flashlight with an integrated filter placed over the lens to prevent the subject from seeing the source. This provides a working range of up to 50 feet (reliable) and can be increased to several hundred using a more powerful source such as a 5- to 6-cell flashlight. Needless to say, the builder has total flexibility in choosing his IR source and can adjust the optics to meet his needs.

Example: Long range quick viewing may utilize a small 2-cell light with eight nicad AA cells to replace the normal two D-cells providing a significantly brighter IR source yet lasting for less time than the normal D cell would. The unit can also be operated using external sources such as super intense filtered automobile headlamps extending the range out to four to five hundred feet and providing a wide field of illumination. Note that viewing of active IR sources such as lasers etc., do not require the internal IR source.

THEORY OF OPERATION

A subminiature high voltage power supply described in Figs. 3-1 through 3-3 produces approximately 15 kV at several hundred microamperes from a 7- to 9-volt rechargeable nicad battery. This voltage is applied to the tube (TUB1) Fig. 3-4 with "plus" going to the viewing end and "negative" to the objective end. A voltage dividing network consisting of resistor R8, R9 and R10 provide electronic focusing via the grid ring.

An objective lens (LE1) with adjustable focal length gathers the reflected image, illuminated by the IR, and focuses this image at the objective end of the tube. Image conversion now takes place inside the tube and is displayed on the viewing screen of the tube in a greenish tinge. Viewing resolution is usually adequate to provide subject identification at distances to 50 feet or more (depending on the intensity of the IR source and quality of optics).

CONSTRUCTION STEPS

1. Assemble and test power board as outlined later in this chapter. Note Figs. 3-1, 3-2, and 3-3. Check for absence of corona in high voltage section. Remove all sharp points and insulate with corona dope etc. See Table 3-1.

2. Solder R9 between "focus ring" terminal and viewing screen ring. This must be done quickly with at least a 40-watt soldering iron as the possible "glass to metal" seal may be damaged if allowed to heat up. Note that the resistor is dressed as shown in Fig. 3-4. The +HV lead from the power board is connected to the remaining lead of R9.

3. Solder R8 between "objective ring" and "focus ring". Note that the resistor may be a combination of one or several and it may be wise just to solder a short piece of buss wire to the objective ring as a connection point for these possible combinations along with the negative HV lead from the power board. Note the different combina-

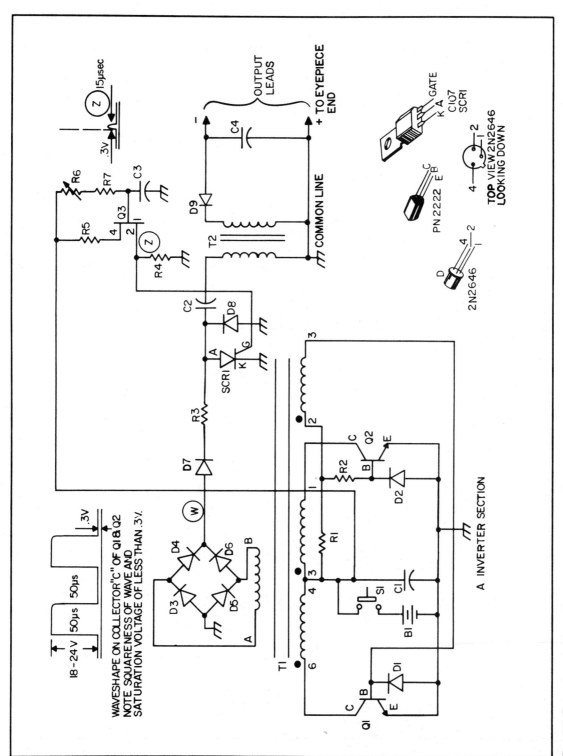

Fig. 3-1. Power board schematic.

41

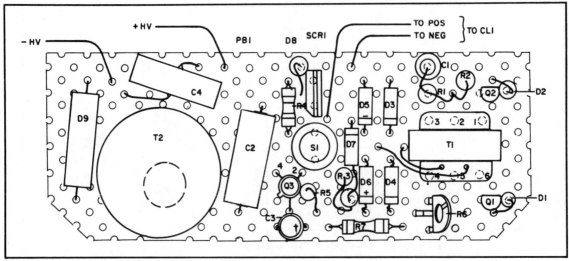

Fig. 3-2. Power board component location.

tions of resistors that may be required. See Fig. 3-4 showing R8 combination and solder points. **Caution! Do not solder near the glass seal!**

    4. Obtain some window screen and place it flush against the objective end of the image tube TUB1. Secure the tube on the bench via modeling clay etc., and temporarily connect to the leads from the power board. Observe proper clearance of leads and components. Note the tube glowing greenish and an image of the screen appearing either sharp or blurred. If image is good and sharp you are in luck. If not resistor R8 must be trimmed by either adding or reducing until the image of the screen is sharp.

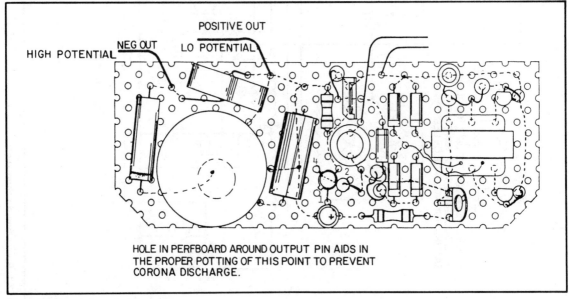

HOLE IN PERFBOARD AROUND OUTPUT PIN AIDS IN
THE PROPER POTTING OF THIS POINT TO PREVENT
CORONA DISCHARGE.

Fig. 3-2A. Power board wiring aid.

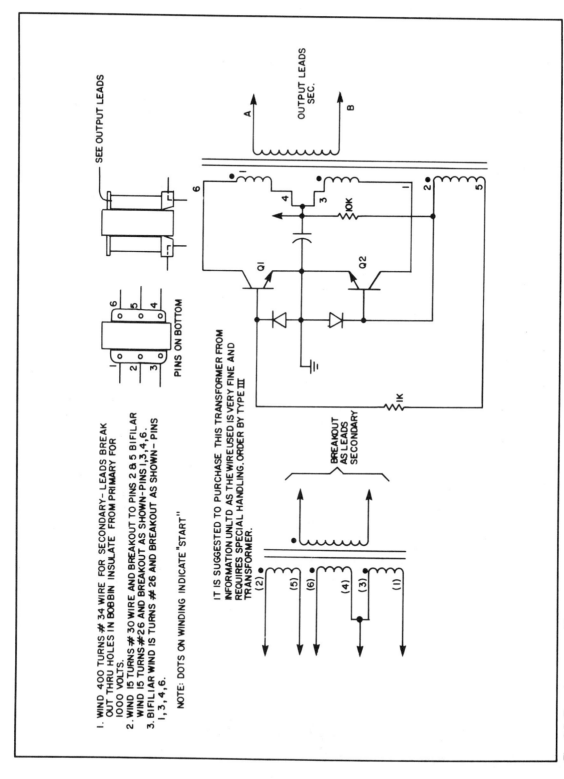

1. WIND 400 TURNS # 34 WIRE FOR SECONDARY– LEADS BREAK
   OUT THRU HOLES IN BOBBIN INSULATE FROM PRIMARY FOR
   1000 VOLTS.
2. WIND 15 TURNS # 30 WIRE AND BREAKOUT TO PINS 2 & 5 BIFILAR
   WIND 15 TURNS #26 AND BREAKOUT AS SHOWN–PINS 1,3,4,6.
3. BIFILIAR WIND IS TURNS # 26 AND BREAKOUT AS SHOWN – PINS
   1,3,4,6.

NOTE: DOTS ON WINDING INDICATE "START"

IT IS SUGGESTED TO PURCHASE THIS TRANSFORMER FROM
INFORMATION UNLTD AS THE WIRE USED IS VERY FINE AND
REQUIRES SPECIAL HANDLING. ORDER BY TYPE III
TRANSFORMER.

SEE OUTPUT LEADS

6 5 4

2 3

PINS ON BOTTOM

OUTPUT LEADS
SEC.

A

B

10K

Q1

Q2

1K

BREAKOUT
AS LEADS
SECONDARY

(2)

(5)

(6)

(4)

(3)

(1)

Fig. 3-3. Transformer type III.

43

Table 3-1. Parts List (SD5).

*R9	(2) 2	(2) 100 MΩ (2) 22 MΩ
*R8	20	(20) 100 MΩ for 2000 MΩ
HA1	1	8" length 1-1/2" ID sked 40 gray PVC
TU1	1	3-1/2 length × 2" ID sked 40 gray PVC
TU2	1	1-1/4 × 2" sked 40 for eyepiece
		2-1/2 × 2" sked 40 for 29 × 43 lens
EN1	1	7" length × 2-1/2" ID sked 40 gray PVC
CA1	1	Lens retaining cap 2" hole (FAB) from 2-3/8" plastic cap
CA2,5	2	Shim-remove end section (FAB) from 2-3/8" plastic cap
CA3	1	Front tube retainer 1-1/4" hole (FAB) from 2-3/8" plastic cap
CA4	1	Rear tube retainer 15/16" hole (FAB) from 2-3/8" plastic cap
CA7	1	Eyepiece retainer (FAB) from 2-3/8" plastic cap
SH1	1	3/4" length 2" ID gray sked 40 PVC (FAB)
CA6	1	1-7/8" plastic cap
LE1	1	54 × 89 objective lens
LE2	1	29 × 43 lens if eyepiece not available
MEM1	1	Rubber membrane-use section bike tire (FAB)
BRK1,2	2	9" × 1/2" aluminum strip (FAB)
SW1	4	6-1/4" sheet metal screws
SW2	4	1/4" 20-1 nylon screws
IRF1	1	Small 43 mm filter IR
*PBK3K	1	PBK3K Kit & plans power board Table 3-2

Special Parts and Assemblies

*TUB1	6032 Image Tube (Please note these are in scarce supply at the time of this writing). Write or call for price & availability.
*PBK30	Assembled power board ready to use.
*RES KIT	Single 2000 MΩ with (2) 100 MΩ and (2) 22 MΩ greatly simplifies assembly.
*WAF1	50 mm wide angle lens
*TEL1	75 mm telephoto lens
	(These lenses are necessary for actual viewing of scenes etc.)
*CMT1	"C" mount adapter for above lenses
*HLR1	High powered IR source for long range viewing

Most parts and assemblies available from Information Unlimited, P.O. Box 716, Amherst, NH 03031. Write or call (603) 673-4730 for price and availability. Parts marked with an asterisk are available individually.

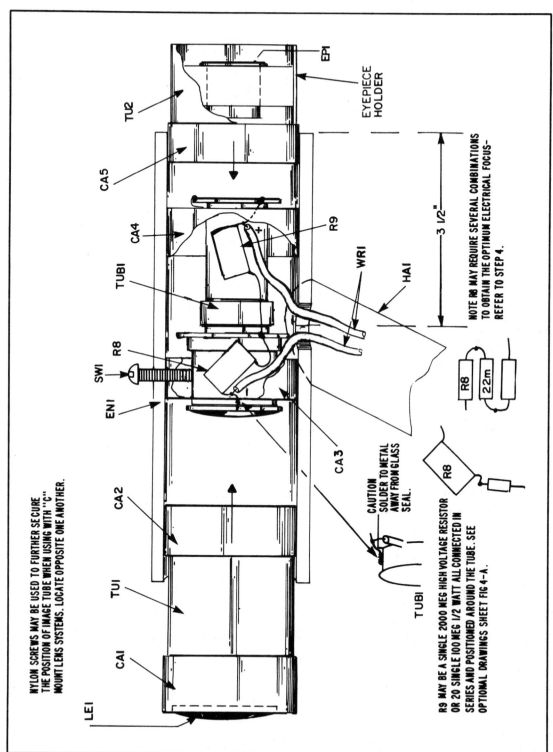

NYLON SCREWS MAY BE USED TO FURTHER SECURE
THE POSITION OF IMAGE TUBE WHEN USING WITH "C"
MOUNT LENS SYSTEMS. LOCATE OPPOSITE ONE ANOTHER.

LEI

CAI

TUI

CA2

ENI

SWI

R8

TUBI

CA4

CA5

TU2

EPI

EYEPIECE
HOLDER

R9

WRI

HAI

3 1/2"

NOTE R8 MAY MAY REQUIRE SEVERAL COMBINATIONS
TO OBTAIN THE OPTIMUM ELECTRICAL FOCUS—
REFER TO STEP 4.

R8

22m

CA3

CAUTION
SOLDER TO METAL
AWAY FROM GLASS
SEAL.

R8

R9 MAY BE A SINGLE 2000 MEG HIGH VOLTAGE RESISTOR
OR 20 SINGLE 100 MEG 1/2 WATT ALL CONNECTED IN
SERIES AND POSITIONED AROUND THE TUBE. SEE
OPTIONAL DRAWINGS SHEET FIG 4-A.

TUBI

Fig. 3-4. X-ray of optics & tube.

45

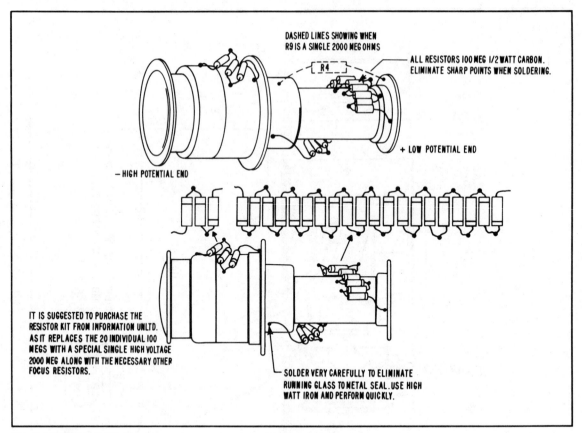

Fig. 3-4A. Optional drawing.

This step electro statically focuses the system. Our lab values were 2000 meg for R9 and approximately 200 meg for R8. You may use a combination of 100 and 22 meg 1/2-watt carbon resistors or the single 200-meg HV series depending on the value required for sharpest focus. You may now use RTV to further secure the resistors and insulate the HV points against possible corona.

5. Position resistors and slip on CA3 and CA4 position bushings. These are made from 2-3/8 inch plastic caps. CA3 is fabricated by carefully removing a 1-1/4 inch circle. CA4 is fabricated by carefully removing a 1-inch circle. This must be done as perfectly centered as possible. Note HV wires snaking between tube body and bushings.

6. Fabricate EN1 from a 7-1/2 foot length of 2-3/8 inch ID schedule 40 PVC tubing. Note hole adjacent to HA1 handle for feed of HV wires to tube from power board.

7. Fabricate HA1 handle from an 8 inch length of 1-1/2 inch ID schedule 40 PVC tubing. Note access hole for S1 activation. Tube must be shaped and fitted where it abuts to EN1 main enclosure. See Figs. 3-5 and 3-6.

8. Fabricate BRK 1,2 brackets from a 1/2-inch wide strip of 22 gauge aluminum as shown Fig. 3-7. Note holes for #6 × 1/4 sheet metal screws for securing assembly together.

9. Fabricate TU2 from a 3-3/4 inch length of 2-inch ID schedule 40 PVC tubing

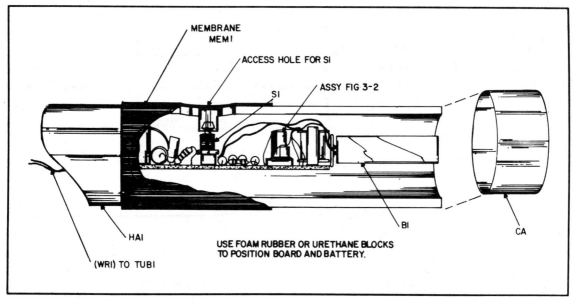

Fig. 3-5. X-ray view showing power board and battery in handle.

for the objective lens. Note this is only 2 inches long when using the optional optics and "C or T" mount adapter fitting.

10. Fabricate TU2 from a 1-1/2 inch length of 2-inch OD schedule 40 PVC tubing for the eyepiece.

11. Fabricate LRET1 eyepiece retainer from a 1/2 inch thick piece of plastic or wood, etc. This piece must fit snugly into TU2 as shown. Eyepiece EP1 is now snugly fitted into cutout in LRET1. This assembly may vary according to the eyepiece lens or assembly used.

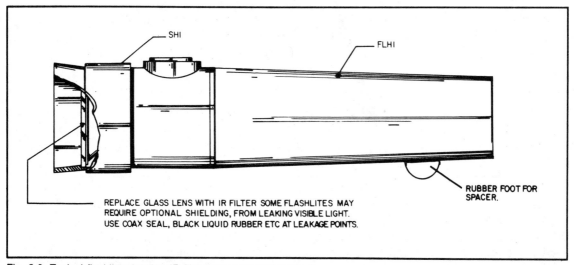

Fig. 3-6. Typical flashlite showing IR filter and mounting method.

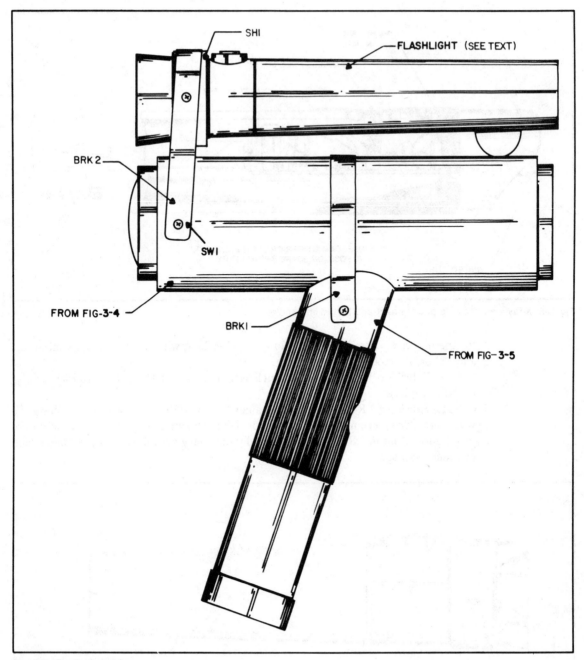

Fig. 3-7. Final assembly.

12. In order for TU1 and TU2 to telescope into the main enclosure EN1, suitable cylinderical shims CA2, CA5 must be fabbed. These can be the 2-3/8 inch plastic cap similar to that used for CA3 and CA4. They now have their ends removed so as to be shim rings as shown in Fig. 3-4. This method is cheap and works reasonably well.

The builder obviously could substitute the machining of properly fitted parts if he desired, fabbed from aluminum, plastic etc. This approach is more professional looking but can be much more costly.

13. Fabricate CA1 by placing over TU1 and cutting out center using wall of tubing as a guide for knife. Lens LE1 is now retained by action of CA1 securing to end of TU1. See Fig. 3-8.

14. The lens shown used is a simple uncorrected convex that is adequate for most IR source viewing. It is not a viewing lens such as the optional 50mm wide angle or 75mm telephoto with the "C" mount threads. When using these lenses it is suggested to either fabricate or purchase an adapter ring that will adapt to lens threads and fit snugly into the enclosure. See (CMT1) Fig. 3-9.

15. Disconnect the wires from the power board and snake them up the access hole in EN1 as shown. Slide in TUB1 image tube and note position of CA3 and CA4 position bushings securing and centering it in place as shown in Fig. 3-4. Be careful not to bunch or bind wires. Resolder leads to power board. Check polarity.

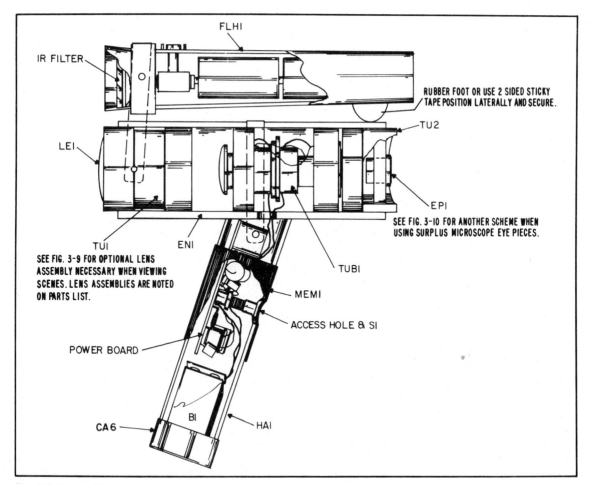

Fig. 3-8. X-ray of final assembly.

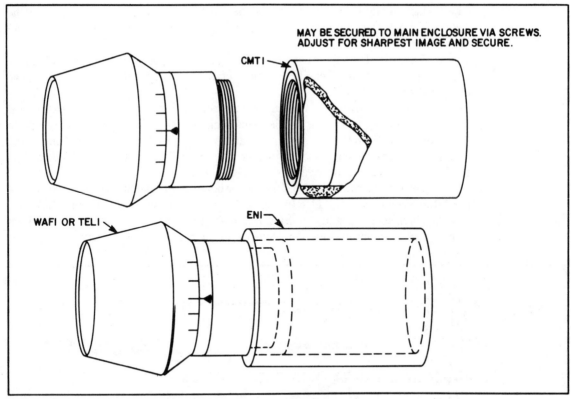

MAY BE SECURED TO MAIN ENCLOSURE VIA SCREWS. ADJUST FOR SHARPEST IMAGE AND SECURE.

CMT I

WAFI OR TEL I

ENI

Fig. 3-9. Optional objective lens drawing.

16. Insert the power board into the HA1 handle. The wires should be long enough for complete removal of assembly when handle is secured in place via BRK1 bracket. This allows any preliminary adjustment. Leads may be shortened once proper operation is verified. Connect battery to power board and energize switch S1. Adjust trimmer R6 for a reliable unflickering image (screen still in place) on the face of the tube. R6 must be set on the low current drain end. Check for any possible readjusting of focusing voltage. If you did your homework you will not have to readjust divider values. Once operation is verified, check for any excessive corona and elimiate. Position board to switch S1 adjacent to access hole in handle (Fig. 3-5). It may be necessary to further secure in place via foam rubber pieces, RTV etc. Slide membrane over access hole and insert battery and cap CA1.

17. Insert eyepiece ad objective lens assembly and check for focusing and imaging of eyepiece. Use a source of low light for illuminating various objects if not in darkness for this step.

18. The unit is shown with a built in IR source consisting of a common 2-cell flashlight fitted with a special IR filter (Fig. 3-6). Any visible light leaks must be sealed with either electrician's gunk, coax seal, or black liquid rubber.

This approach allows total flexibility in viewing sources not requiring IR illumination as the light need not be energized or even removed. The light source may also be intensified by replacing the two D cells with an 8-cell AA nicad pack providing ap-

50

proximately 9 volts. A suitable lamp may be substituted providing several times more illumination. Lamp and battery life will be greatly reduced as this approach is only intended for intermittent use. Note the now available Halogen lamps are far more intense and make excellent IR sources.

Longer range viewing may be accomplished by using other more intense sources such as higher powered lights, auto headlamps, etc. These must be fitted with the proper filters to be usable. Range of several hundred meters may be possible with these higher powered sources. A source capable of allowing viewing out to 500 feet is referenced in the parts list.

To obtain maximum performance and range from the system may require the optional lens system specified. Viewing of externally illuminated IR source will not require the integral IR source.

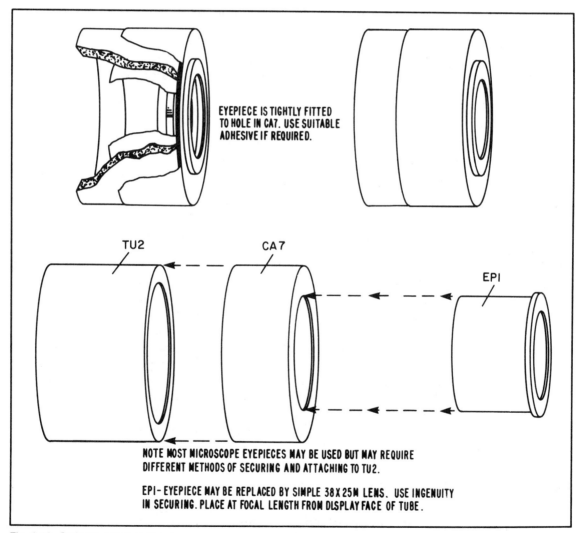

EYEPIECE IS TIGHTLY FITTED TO HOLE IN CA7. USE SUITABLE ADHESIVE IF REQUIRED.

TU2 CA7 EPI

NOTE MOST MICROSCOPE EYEPIECES MAY BE USED BUT MAY REQUIRE DIFFERENT METHODS OF SECURING AND ATTACHING TO TU2.

EPI- EYEPIECE MAY BE REPLACED BY SIMPLE 38 X 25M LENS. USE INGENUITY IN SECURING. PLACE AT FOCAL LENGTH FROM DISPLAY FACE OF TUBE.

Fig. 3-10. Optional eyepiece drawing.

You will note that this device is excellent for viewing the output of our SSL3 Solid-State Gallium Arsenide Injection Laser System, CWL1 Continuous-Wave Solid-State Laser and any other source of infrared energy in the 9000 Å spectrum. No internal IR source is necessary when viewing these actual sources. See Figs. 3-9 and 3-10.

POWER BOARD (PBK3) THEORY OF OPERATION

The Inverter Section consists of switching transistors (Q1) and (Q2) that alternately switches the primary windings of a saturable core transformer (T1). A high-voltage square wave is induced in the secondary of T1 via this switching action and is rectified by diode bridge D3, D4, D5, and D6. Base current drive for Q1 and Q2 is obtained by a tertiary feedback winding on T1 and is applied in the correct phase to turn the appropriate transistor on. This base current is limited by resistor R2. Diodes D1 and D2 provide a return path for the base current flowing in the opposite transistor respectively. R1 serves to unbalance the circuit to initiate switching. A voltage of approximately 200-300 volts is obtained in this circuit from a 7- to 9-volt nicad battery pack (B1). Higher powered operation may be obtained by increasing B1 to a 12-volt battery pack, however, more space is required and care must be taken not to overrate the components if continued use is anticipated. See Figs. 3-1 to 3-3.

The Capacitor Discharge Section consists of a high-voltage pulse transformer T2, Figs. 3-1 and 3-2, being current pulsed via SCR1 shorting a charged capacitor C2 across its primary. C2 and the primary inductance of T2 provides a ringing wave whose negative overshoot commutates SCR1 to turn off. It is important that this primary inductance be sufficient so when combined with capacitor C2 allows a ringing frequency with a period considerably larger than the required commutation turn off time of SCR1. Diode D8 provides energy recovery of the negative overshoot component of this discharge pulse.

Transformer T2 now force induces a very high voltage pulse in its secondary with a high instantaneous peak current (this system is similar to a Capacitor Discharge Ignition). Diode D7 and R3 limit the dc current to the SCR1 and prevents dc lock on, while also providing a high impedance to the negative turn off pulse.

SCR1 is triggered by the UJT pulse timing circuit consisting of Q3. Pulse repetition rate is determined by capacitor C3 and the charging resistor R6. SCR1 switch rate can be adjusted "from one to ten pps." Higher pulse rep rates may have a tendency to overload the inverter power supply, where it will be unable to supply the current necessary to successfully charge C2, consequently with its charge voltage dropping off.

The voltage output of T2 is well over 10 V and is rectified thru D9 and integrated on C4 providing a dc voltage of 10-12 kV. You will note that the diode is supplying negative voltage at its high potential in reference to power supply common. This helps prevent any high voltage occurring adjacent to the eyepiece as now the objective end is above common.

ASSEMBLY OF POWER BOARD

1. Layout and identify all parts and pieces.
2. Assemble T1 as per Fig. 3-3, type III transformer instruction sheet or obtain an assembled and tested unit from Information Unlimited. See parts list (Table 3-2).

Table 3-2. Power Board Parts List (PBK3).

R1,3	2	10 kΩ 1/4-watt resistor R3 use two in parallel
R2,8	2	1 kΩ 1/4-watt resistor
R4,5	2	100 Ω 1/4 watt resistor
R6	1	100 kΩ trimpot 1/4 watt resistor
R7	1	2.2 kΩ 1/4 watt resistor
C1	1	10 μF/25V capacitor elect.
C2	1	.1 μF/400 capacitor disc
C3	1	1 μF/25V capacitor elect
C4	1	.001 @15 kV ceramic capacitor HV
Q1,2	1	PN2222 transistor NPN
Q3	1	2N2646 UJT
D1,D2	2	1N4002 diode 100 V
D3,4,5 6,7,8	6	1N4007 diode 1000 V
*D9	1	20 kV diode
SCR1	1	C107D SCR
*T1	1	Inverter transformer, Info. Unltd. type III.
*T2	1	Pulse transformer special, Info. Unltd.
*PB1	1	Perfboard 3-3/4″ × 1-1/2″ (Printed circuit board available)
S1	1	Push button switch
CL1	1	Battery clip (heavy duty)
WR1	24″	HV leads 10 kV
B1	1	7.2 volt nicad (not included)

Special Parts and Assemblies

PBK3	Power Board Parts Kit
*PBK30	Assembled Power Board
*PCSD5	Printed Circuit Board
*BC9K	Battery Charger Kit
*BC90	Battery Charger Assembled

Most parts and assemblies available from Information Unlimited, P.O. Box 716, Amherst, N.H. 03031. Write or call (603) 673-4730 for price and availability. Parts marked with an asterisk (*) are individually available.

3. Fabricate PB1 as shown from a 3.8 inch × 1.4 inch piece of perfboard or use optional available PC board. Note enlarged holes for pushbutton switch S1.

4. Insert and wire components as shown. Avoid wire bridges and use component leads wherever possible. Follow layout in Fig. 3-2. Observe polarity of diodes, semiconductors and capacitors. Always leave at least 1/16 inch to 1/8 inch leads on semiconductors so they do not sit directly on board. *Note: This power board assembly is used in several of our devices with minor revisions such as S1 and R6 being externally mounted via interconnecting leads.*

5. Attach and wire T1 transformer being very careful not to break the thin wires of the output winding. Use RTV to attach transformers when using perfboard. When T1 is purchased from Information Unlimited it will have pins for mating with the optional printed circuit board. However, the secondary output circuitry will still be the wire leads that must be secured and held in place with RTV or equivalent. Do not connect *secondary* of T2 high voltage pulse transformer at this time.

6. Connect any external leads and battery clip CL1. Verify wiring accuracy and quality of solder joints. Check for shorts or danger points.

7. Connect a 6-volt source to CL1 and measure approximately *175 volts* at point W (Fig. 3-1). Input current should be between 50 and 75 mA.

You may want to check the waveshapes on the collectors of Q1 and Q2 and point Z. This is usually not necessary if the above measurements are verified. Finger touch Q1 and Q2 and check for cool or slightly warm.

8. You may connect in T2, D9, and C4. *These connections must be shiny and properly spaced or flashover will occur.* Form a temporary gap of no more than 1/4 inch at output leads. This prevents damaging overvoltage from being produced across C4. Note that a large hole is shown in the perfboard around the output pin of T2. This allows the potting wax or compound to better seal this point.

9. Reconnect the 6-volts supply and immediately note a sparking across the temporary gap. Quickly adjust R6 to limit and note sparking rate changing accordingly. Note values of input current at these limits. Readjust to lowest rate as this setting demands the minimum amount of power.

10. Connect appropriate output leads.

11. Carefully clean HV output connections and verify absence of leakage by observing operation in the dark. Prepare some hot paraffin wax and dip board up to only the HV components consisting of T2, D9, and C4. Repeat and build up a healthy layer. Several layers of varnish may also be used in place of the wax. Power board is ready for action!

Chapter 4

LASER RADIATION –
AVOID DIRECT EXPOSURE
TO BEAM

MAX POWER .014 W
wavelength 830 nm

CLASS IIIb LASER PRODUCT

Continuous-Wave
Solid-State Lasers

T HIS PROJECT SHOWS HOW TO UTILIZE A RELATIVELY NEW DEVICE THAT HAS been available only in the last several years. Chapter 1 describes a solid-state pulsed IR laser that can generate high peak powers in the tens of watts, however, the duration and repetitive rate limits the energy down into the tens of milliwatts. The continuous-wave (cw) solid-state laser is able to produce equivalent amounts of energy without the need for fast high-current pulses. Semiconductor junctions are now able to generate lasing using currents in the magnitude of several hundred milliamps, thus allowing continuous energy output. The drawbacks of these devices are extreme sensitivity to transients, noise, temperature, application of over current and, of course, high cost. Special power supplies are required for these devices and they are also costly.

The general theory is covered in Chapter 1 with the main difference being that lasing occurs at a far less threshold of current and, therefore, operation becomes continuous without the use of large heat sinks, etc.

The circuit we describe here (Fig. 4-1 and Table 4-1) is only a suggestion for the experimenter and should not be considered a product for resale in kit or completed form. Its main objective is to supply power to these devices with a minimum of cost and complexity. Spec sheets are shown here for two of these devices, compliments of M/A-COM Laser Device Labs (Fig. 4-2). They can be purchased through them along with the intended power supplies. You will note the labelling of these devices as specified by the BRH are Class IIIB and therefore require all compliances as listed in the Laser Safety Section of this book (see pages 7 through 12).

The circuit described utilizes the relative voltage stability that a nickel-cadmium battery supplies while able to deliver a wide range of current requirements. Figure 4-1

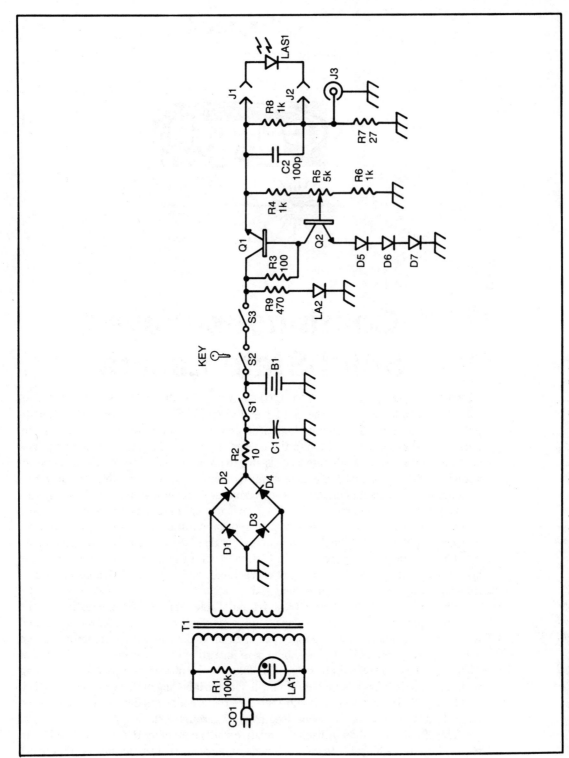

Fig. 4-1. Suggested experimental power supply for continuous wave laser diode.

Table 4-1. Experimental Power Supply Parts.

R1	1	100 kΩ 1/4-watt resistor
*R2	1	10 kΩ 1/2-watt precision resistor
*R3	1	100 Ω 1/2-watt precision resistor
*R4,6,8	3	1 kΩ 1/4-watt precision resistor
*R5/S3	1	5 kΩ pot/switch
*R7	1	27 Ω 5-watt precision resistor
R9	1	470 Ω 1/4-watt resistor
LAS1	1	See Fig. 4-2
C1	1	100 μF 25 V elect capacitor
C2	1	100 pF 50 V disc
D1,2,3,4 5,6,7	7	1N4002 1 amp diodes
Q1	1	2N3055 NPN pwr T03 transistor
Q2	1	D40D5 NPN pwr tab transistor
T1	1	12 V at 100 mA power transformer
LA1	1	Neon lamp
LA2	1	Led FLV 106
CO1	1	Power cord
S1	1	SPST switch "charge"
S2	1	Key switch nonremovable key in "on" position
J1,2	2	Terminal or binding posts
J3	1	RCA phono jack
TO3	1	Mounting kit for Q1
PB1	1	Small piece of perforated circuit board
EN1	1	Small aluminum enclosure
B1	8	10 V battery - Use (8) 1.2 volt AA Nicads
BH1	1	8 AA cell holder
		Miscellaneous hardware, wire, bushing etc.
		Proper labels as shown Fig. 1-8.

(Parts marked with an asterisk should be low temperature coefficient resistors.)

All the parts shown in Fig. 4-1 are available except the laser diode through Information Unlimited, P.O. Box 716. Amherst, NH 03031. Write or call (603) 673-4730.

For the serious experimenter or laboratory technician, it is suggested to consider the special current power supply model LCS-350/R available through M/A-Com Laser Diode. See Fig. 4-2.

shows the schematic of a charging system plus a means of controlling the required current to these laser diodes. An external digital millimeter is required for precise control. Note that this circuit has no provisions for thermal drift except direct monitoring of the meter to prevent current run away.

Before the power supply is used, the Nicad batter (B1) consisting of eight AA cells must be fully charged by activating charge switch (S1). Indicator lamp (LA1) ignites when primary line power is applied to step-down transformer (T1). A full-wave rectifier consisting of diodes (D1 through D4) supplies positive going pulses to capacitor (C1) through current-limiting resistor (R2). B1 now fully charges at approximately a 50 mA rate. Primary power is now removed isolating the circuitry from the 115-volt ac line via S1. Operate switch (S2) is open with voltage control pot (R5) set at minimum output. The monitoring current meter is securely connected to jack (J3). Note this meter is actually reading a voltage across the laser diode limit resistor (R7) and must be converted to milliamps by using Ohm's law. It is a good idea to initially make a calibration chart showing "volts read" = "milliamps" to laser diode. The diode is attached to a heatsink and then connected to terminals (J1 and J2) making a doublecheck

M/A-COM LASER DIODE, INC.

TYPE LCW-10
TYPE LCW-10F
with FIBER PIGTAIL
MULTIMODE
GaAlAs CW
INJECTION LASER DIODE

ACTUAL SIZE

FEATURING:
- CW OPERATION
- UP TO 14 mW OUTPUT POWER ON SELECTED UNITS
- CHOICE OF WAVELENGTH
- HIGH TEMPERATURE OPERATION
- LOW THRESHOLD CURRENT

DESCRIPTION: The LCW-10 is a multimode
CW GaAlAs double heterostructure laser diode.
This device features low threshold current and high output power.
The diodes are passivated for long life and reliability.
They offer good performance at elevated temperatures.
The LCW-10F is the fiber coupled version of the LCW-10.
This unit is available with standard fiber as noted below.
Custom fiber attachment available upon request.

Pom. I_{th} & I_o Information furnished with each device — Data at 27 C unless otherwise noted.

		Symbol	Min.	Typ.	Max.	Units
Total Radiant	LCW-10	Pom	4	7		mW
Flux at Rated I_f	LCW-10F*	Pom	1	2.0		mW
Peak Wavelength		λ_p	See Below	830	See Below	nm
Spectral Width @ 50% points		$\Delta\lambda$		2.5		nm
Source Size				0.2x7.0		μm
Rise Time of Radiant Flux		T_r		100	800	ps
Far Field Beam Divergence (LCW-10) @ 50% Points		$\theta_p x \theta_n$		10x35		degrees
Threshold Current		I_{th}		90	150	mA
Operating Current		I_o		115	I_{th} + 50	mA
Differential Quantum Efficiency				0.3		mW/mA
Forward Voltage at I_f		V_o		2.0		volts
Operating Temperature		T_o	0		60**	°C
Storage Temperature		T_s	−55		125	°C

*Standard fiber pigtail is all glass. 50μm graded index core. 125 μm O.D. with N.A. .20
**Selections to higher operating temperatures are available

WAVELENGTH SELECTION			ABSOLUTE MAXIMUM RATINGS (CW)

WAVELENGTH SELECTION

OPTION	WAVELENGTH RANGE
A	800 to 810 nm
B	810 to 840 nm
C	840 to 860 nm
D	860 to 880 nm

ABSOLUTE MAXIMUM RATINGS (CW)

Maximum Forward Current = lower of I_{th} + 60 or 210 mA
Maximum Reverse Voltage = 2.0 volt
Maximum Operating Temperature = 70°C
Maximum Storage Temperature = 140°C
Minimum Storage Temperature = −75°C
Maximum Radiant Output Power = 14 mW

Fig. 4-2. The LCs-350/R dc current source (Courtesy of M/A-COM Laser Diode, Inc.). (Continued through page 60.)

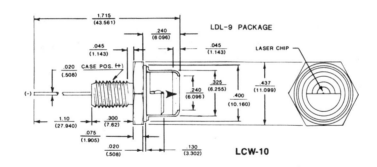

LDL-9 PACKAGE

LCW-10

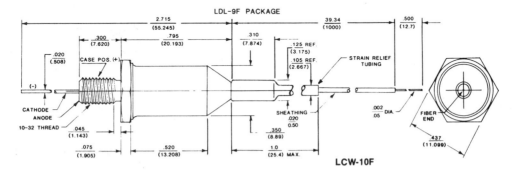

LDL-9F PACKAGE

LCW-10F

LASER SAFETY

Gallium Arsenide lasers emit infrared radiation from the glass window on the top of the block package which is invisible to the human eye. When in use, safety precautions should be taken to avoid the possibility of eye damage.

Do not stare directly at the device or view an operating laser at close range. If viewing is required, the beam should only be observed by reflection from a matte surface utilizing an image convertor or by use of a suitable fluorescent screen.

CAUTION

Use of controls or adjustments or performance of procedures other than those specified herein may result in hazardous radiation exposure.

M/A-COM LASER DIODE, INC.
1130 SOMERSET STREET
NEW BRUNSWICK, NJ 08901
(201) 249-7000
TWX 710-998-0597

LCS-350/R

The LCS-350/R DC current source is specifically designed to provide extremely stable, transient-free output and is ideally suited for driving a wide variety of CW injection laser diodes (ILD's), light emitting diodes (LED's), and other precision electronic components. The LCS-350/R consists of a regulated power supply, a critically damped ramp generator, and a low-noise, non-capacitive output stage with associated protection circuitry. The unit contains a precision 10-turn linear potentiometer which permits the output current to be manually adjusted to any level up to 350mA with an accuracy of 1mA as displayed on the built-in digital meter. The ramp generator function is incorporated to allow a slow sweep up to any preset current within the 350mA range with no measurable overshoot. Together with the 1mV/mA X-Y plotter output, the ramp function provides a convenient and simple method for driving ILD's and LED's when measuring characteristic optical power output versus input current (P-I) curves. Also included is an open circuit warning lamp which, when activated, indicates the presence of an excessive potential across the output terminals. This feature, along with the LCS-350/R's transient suppression network, virtually eliminates any possibility of accidental damage to sensitive electro-optic components when used in accordance with the instructions given in the unit's detailed operating manual.

ELECTRICAL

- Output Current : 0 to 350mA
- Compliance Voltage : >6 Volts · 350mA
- Regulation : ±2% ±1 Digit
- Noise and Ripple : <0.01% Typical
- Transients : None
- Stability : ±1mA over 8 Hours
- Rate of Ramp : 7mA/second ±10%
- Power : 105 to 125 VAC · 47-420 Hz (Standard)*
- Output to X-Y Recorder : 1mV/mA ±1%
    * Option: 210 to 250 VAC · 47-420 Hz.

MECHANICAL

- Dimensions : 8-1/4" W x 4-1/2" H x 6-1/2" D
- Connections : Current output - Five-way binding posts
  : X-Y plotter output - Amphenol #126-196 seven pin socket supplied with matching plug.
- Line Cord : 6 Ft. with three prong grounded plug.
- Operating Temp. : 10 to 40 C
- Storage Temp. : 0 to 55 C

DISPLAY

- 3-1/2 Digit LED

M/A-COM LASER DIODE, INC.

on polarity, because a mistake here will result in disaster. You will note resistor and capacitor combination (R8, C2) connected between J1 and J2 to discharge any potentially damaging voltages or fast transients.

Power-up switch (S2) is now energized with light emitting diode (LA2) indicating "power on." An infrared indicator such as that described in Chapter 1 is placed adjacent to the output of the laser. R5 is now slowly rotated noting the current as indicated by the monitoring meter. Current is adjusted to a value just over the threshold.

The method of voltage control is nothing more than current-pass transistor (Q1) being controlled by transistor (Q2). The variable voltage output is fed to the laser diode through current resistor R7. R7 supplies a form of ballast and also serves as a means of developing the monitoring voltage for the meter. Current may be set to the rating of the diode connected, however, it is suggested to stay below this value, as the questionability of the monitoring meter accuracy combined with the tendency of the device to draw more current as it heats up makes this a wise approach.

The experimenter may now wish to have the diode mounted such as that in Chapter 1, Fig. 1-5 showing the methods of heat sinking, Fig. 1-7 showing some basic optics and Fig. 1-8 showing labelling as required by the BRH (Bureau of Radiological Health and Welfare). This now constitutes a two part laser system consisting of the power supply connected to the laser diode assembly via a cable, etc. A label noting a noninterlocked housing is now required along with the other compliances as listed in the Laser Safety Section of this book (pages 7 through 12).

Power down is simply the reverse of powering up and must be followed or irreversible damage may be done to the laser diode. For the experimenter wishing to further work with the devices as far as modulation temperature stabilizing etc., may wish to consult current literature available from M/A-Com or Hamamatsu.

Chapter 5

Fiberoptic Communication

T HE INFPRMATION ON PAGES 64 THROUGH 77 IS COURTESY OF ITT CANNON. IT is presented as a full study of fiberoptics theory and related data. Several sources of kits and associated materials are referenced at the end of the chapter.

The following related material is available from Information Unlimited, P. O. Box 716, Amherst, N.H. 03031. Call (603) 673-4730 for pricing and availability.

FIBK1—Fiber Optic Kit intended for the experimenter who wishes to produce interesting visual effects and phenomenon.

FBK1—Fiber Lighting manual for the hobbyist and craftsman, fully illustrated.

1 INTRODUCTION

Fiber optics communication is a technology resulting from the successful integration of:-

a) optical communication—the transmission of information as a modulated beam of light, and

b) fiber optics—the guidance of light within flexible filaments often no thicker than a human hair.

Both these technologies have enjoyed a limited following in their separate fields, but it has been the additional benefits provided by their combination that have attracted worldwide interest, and have led to the establishment of major research and development programmes by several organisations.

Fiber optics communication has now moved out of the laboratory! ITT, for example, offer as a complete package a fully specified link for commercial applications. Whilst this can be successfully installed and operated without specialised knowledge or skill, an understanding of the basic principles is not difficult to acquire: indeed, it is desirable as it leads to a better understanding of both the potential and the limitations of the technology.

A frequently encountered deterrent is the need to extract the relevant facts from the abundance of information available, much of which is either more detailed than required for the non-specialist, or alternatively is too limited in its scope. This publication is aimed to fill the need for a basic fiber optics communication manual, by describing the principles of the technology in a concise but thorough form, whilst avoiding unnecessary mathematical theory.

USE OF MANUAL
Fiber optics users have many differing skills and backgrounds: hence the need for a comprehensive manual.

We have used a logical layout which ranges from fundamentals to aspects of system design and specification. If you are in a hurry, start at section 10 and read the others as you find necessary. Glancing through the illustrations may be the quickest way to find an area of particular interest or need.

HISTORY OF FIBER OPTICS
The principle of guidance of light within a 'transparent conductor' is not new—in 1870 John Tyndall demonstrated to fellow members of the Royal Society that light would follow the curved jet of water issuing from a container.

Later, J. L. Baird filed patents covering the transmission of light in glass rods—intended for use in an early colour television system. However, the practical implementation of this and other proposals was held up by the high intrinsic optical loss inherent in the materials available, and by the absence of an optical cladding.

It was not until the 1950s that fiber optics became a realistic proposition, with many applications being developed, principally devoted to the transmission of visible light, either for remote illumination or for the transfer of images in flexible viewing instruments (e.g. for medical applications).

In 1966 Charles Kao and George Hockham, two scientists working at Standard Telecommunication Laboratories, Harlow, proposed the principle of information transmission via a transparent dielectric medium (e.g. glass fiber). One of the requirements postulated for a viable system was a reduction in fiber attenuation to a target figure of 20 dB/km. With available fibers typically exhibiting 500-1000 dB/km, the necessary advance in materials technology to achieve such a drastic reduction in attenuation appeared hopelessly optimistic, yet it took only four years for this target to be achieved.

Such have been the further advances in the technology that commercial fibers with losses one quarter of the target figure are now commonplace, and laboratory values of less than 1 dB/km have been recorded.

ADVANTAGES OF OPTICAL COMMUNICATION
Although great interest in fiber optics communication has been shown by the telecommunications industry as a possible alternative for long-haul high-bandwidth systems, many simple short distance applications—for example industrial process control, computer installations and high voltage monitoring—are likely to predominate in the near future.

The principal advantages to be gained from using fiber optics are listed below:-

a) **Low loss:-**
   potentially long distances between terminals or repeaters

b) **High bandwidth:-**
   high rates of data transmission

c) **Lightweight/small size:-**
   aircraft/mobile applications

d) **Electrical isolation:-**
   high voltage monitoring or control

e) **Freedom from electro-magnetic interference:-**
   No pickup in electrically noisy environments

f) **No spark hazard:-**
   operation in explosive atmospheres

g) **Security:-**
   very difficult to locate and 'tap'

h) **Open circuit failure mode:-**
   no 'short circuit' fault damage to terminals

2 FUNDAMENTALS OF FIBER OPTICS COMMUNICATION LINKS

FUNCTION

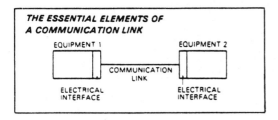

THE ESSENTIAL ELEMENTS OF A COMMUNICATION LINK

A fiber optics link provides an alternative means of communication to wires, twisted pairs, coax or free space radio and microwave systems. The link transmits signals between interfaces in the equipments which are communicating.

ESSENTIAL COMPONENTS

A fibre optics link is composed of a number of basic components:-

a) **Electro-optical signal transducer** — the transmitter

b) **Optical fiber cable**

c) **Opto-electrical signal transducer** — the receiver.

Depending on the particular requirements of the system, the link may also include:-

d) **Demountable connectors**
   - transmitter to fiber
   - fiber to fiber
   - fiber to receiver

e) **Branching couplers**
   - for multiple access systems

The simplest fiber optics links provide *point-to-point* communication, that is, a given transmitter is permanently coupled to a designated receiver, with communication in one direction only. Return communication must be provided by an entirely separate link.

A 'send and receive' *duplex* link must include an optical 'Y' coupler at each end of the communications cable to combine

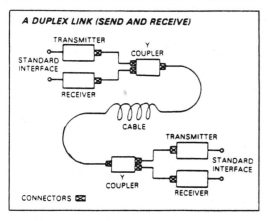

A DUPLEX LINK (SEND AND RECEIVE)

the separate transmitter and receiver connections in each terminal. In optical communications an allowance must be made for the halving of power which occurs in each coupler, but because of the unique isolating property of fiber optics no 'send' and 'receive' switching is necessary.

Cable, connectors and couplers are analogous to their electrical counterparts. These are the interconnection components which permit signal transmission over complex paths between two or more terminals. The design of these components has to satisfy not only the required optical performance, but also mechanical and environmental specifications so that installation and operation will be no more difficult to achieve than with wire cables.

The optical transmitter and receiver are unique to fiber optics. The transmitter usually contains an optical emitter which can either be a light emitting diode or a semiconductor laser, together with associated drive circuits. Similarly the receiver contains a PIN or avalanche photodiode, followed by circuits for signal amplification, output level reconstitution and possibly automatic gain control (AGC).

Signal coding and decoding may also be incorporated to exploit to the maximum the special features of fiber optics transmission.

ANCILLARY EQUIPMENT AND INSTRUMENTATION

Ancillary equipment is needed for:-

a) Fiber termination, and assembly of connectors to cables

b) Permanent jointing or splicing

Techniques must be easy and quick to perform either in the factory or on site, and must result in assemblies of reliable mechanical and environmental performance.

Special purpose instrumentation is needed to characterize the optical path, and for fault location and diagnosis.

SINGLE FIBERS AND FIBER BUNDLES

Two alternative concepts have developed in the manufacture of cables for fiber optics communication, these being the use of either a single fiber or a multiple fiber bundle for each information channel. Historically, the choice has been significantly influenced by the background of the individual manufacturer.

For example, those companies already committed to the manufacture of light guides for other fiber optics applications have readily adapted fiber bundles for communications use. This, combined with the relative ease with which they can be coupled to each other and to sources and detectors, has given bundles an early lead in communications technology.

Conversely, other companies, with no previous experience of fiber optics but with a background of semiconductor technology, have recognised that likely developments in miniature sources and detectors favour the use of the small size transmission medium offered by single fibers. The economics of high quality low loss fiber production reinforces this conviction, while parallel developments in fiber strength and cable design enables single fibers to withstand a considerable degree of mishandling, thus overcoming one of the major objections to the use of single fibers.

Consequently, this manual is essentially dedicated to single fiber communication systems, although many of the principles involved are equally applicable to both.

3   PRINCIPLES OF FIBER OPTICS

REFRACTION AND REFLECTION

> ### *REFRACTIVE INDEX* n OF A MEDIUM
>
> $$= \frac{\text{SPEED OF LIGHT IN A VACUUM}}{\text{SPEED OF LIGHT IN THE MEDIUM}}$$

The path of a ray of light in different materials is influenced by the fact that light travels more slowly in an optically dense medium than it does in a less dense one. A measure of this effect is the *refractive index.*

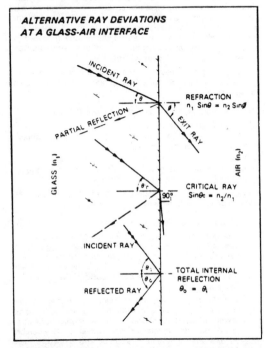

ALTERNATIVE RAY DEVIATIONS AT A GLASS-AIR INTERFACE

It is observed that a ray, approaching an interface (boundary) between media of different refractive indices (e.g. glass-air) at angle θ to the normal in the higher index side, leaves at a greater angle φ in the lower index medium. Snell's law of *refraction* relates these angles by the expression

$$\frac{\sin\theta}{\sin\phi} = \frac{n_2}{n_1}$$

A significant feature of refraction is that a small proportion of the light is reflected back into the originating medium *(partial internal reflection).*

At a particular value of the incident angle θc, the refracted ray emerges parallel to the interface. This is the *critical angle.*

All shallower rays (i.e. θi > θc) are reflected completely, obeying the familiar law of reflection, θo = θi. This is *total internal reflection* (TIR) and is theoretically 100% efficient. In practice, the efficiency can exceed 99.9%, compared for example with 85-90% for a silvered mirror.

TRANSMISSION IN CYLINDRICAL FIBERS

Two parallel glass-air interfaces will trap a ray approaching at a suitably shallow angle, and transmit it, virtually without loss, by a series of total internal reflections.

A cylindrical glass fiber behaves in a similar manner, except that the majority of rays travel in a stepped helical manner by many such reflections at the circumference.

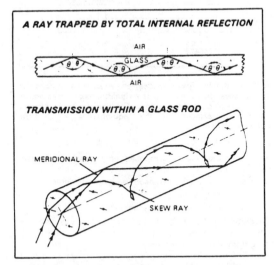

A RAY TRAPPED BY TOTAL INTERNAL REFLECTION

TRANSMISSION WITHIN A GLASS ROD

Since this is difficult to visualise in two dimensions, it is usual to consider only those rays which lie in a plane through the fiber axis *(meridional rays),* even though these are greatly outnumbered by the helical, or *skew rays.*

OPTICAL CLADDING

An essential requirement for total internal reflection is that the external medium should possess a lower refractive index

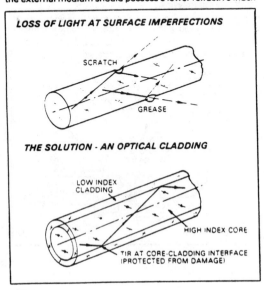

LOSS OF LIGHT AT SURFACE IMPERFECTIONS

THE SOLUTION - AN OPTICAL CLADDING

PRINCIPLES OF FIBER OPTICS

than the fiber. Although this condition is satisfied by a simple glass-air interface, in practice this is easily damaged, and light then escapes by two principal mechanisms, both of which alter the effective angle of incidence at the interface:-

a) inclusions or scratches
b) surface contaminants (e.g. grease)

Grease has a refractive index which is similar to that of glass and hence, to the approaching ray, appears as an extension of the glass surface.

These defects are avoided by the provision of an *optical cladding*. This is a region of lower refractive index surrounding the central zone *(core)*. Total internal reflection can now take place at the protected core-cladding interface.

The development of clad optical fibers represents one of the most significant steps forward in the technology of fibre optics.

NUMERICAL APERTURE

Since only those rays with a sufficiently low grazing angle at the core-cladding interface are transmitted by total internal reflection, rays must approach the input surface of the fiber within a similarly limited cone angle.

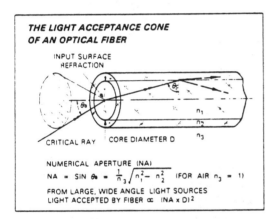

THE LIGHT ACCEPTANCE CONE OF AN OPTICAL FIBER

INPUT SURFACE REFRACTION

CRITICAL RAY CORE DIAMETER D

n_1 n_2 n_3

NUMERICAL APERTURE (NA)

$$NA = SIN\ \theta_a = \frac{1}{n_3}\sqrt{n_1^2 - n_2^2} \quad \text{(FOR AIR } n_3 = 1\text{)}$$

FROM LARGE, WIDE ANGLE LIGHT SOURCES
LIGHT ACCEPTED BY FIBER $\propto$ (NA x D)^2

The half angle θ_a of this cone, within which rays will be transmitted, is called the *acceptance angle* and is related to the refractive indices of the three media—core, cladding and air. The index of air is approximately 1. Note that a refraction occurs at the input surface.

The expression $\sin\theta_a$ is a measure of the light collecting ability of the fiber and is called the *numerical aperture* (NA). Rays at greater angles to the fiber axis than θ_a will not be transmitted by total internal reflection. Since many light sources emit over a wide range of angles, it follows that fibers with large NAs will collect a greater proportion of that light (in proportion to NA^2). Similarly, large diameter fibers are to be preferred if the light source is large compared with the fiber.

Typical fibers have diameters ranging from 30 µm – 600 µm, and NAs from 0.15 to 0.5.

OPTICAL LOSS

The amount of light emerging from the end of a fiber is always less than that entering due to losses caused by scattering and absorption in the core, and by imperfect reflection at the optical interface. This loss is dependent on the length of the fiber, and, like many natural phenomena,

produces an exponential decay, which means that repeated equal increments of length always cause the same proportional decrease in power.

This is an inconvenient law to use for mental calculations, but can be converted into a linear function of length by expressing the exit/input power ratio (Po/Pi) in terms of its logarithm.

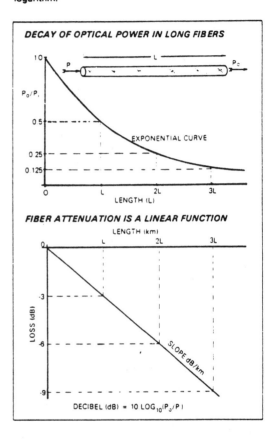

DECAY OF OPTICAL POWER IN LONG FIBERS

P_o/P_i

EXPONENTIAL CURVE

LENGTH (L)

FIBER ATTENUATION IS A LINEAR FUNCTION

LENGTH (km)

LOSS (dB)

SLOPE dB/km

DECIBEL (dB) = 10 LOG$_{10}$(P$_o$/P)

The name *bel* is given to this ratio when using logarithms to the base 10. In practice this is too large for convenience, and a smaller unit one tenth the size has been termed the *decibel* (dB).

Thus, **loss in dB** $= 10 \times \log_{10}(P_o/P_i)$.

Fiber losses are normally expressed in terms of attenuation (in dB) per unit length (km). For example, values range from less than 1 to in excess of 1000 dB/km.

SPECTRAL RESPONSE

Visible light is a small region of the electro-magnetic spectrum covering wavelengths in the range 0.4-0.7 micron. Many of the light sources used in optical communications emit at wavelengths either just inside the visible spectrum (red) or just outside (near infra-red), say from 0.6 to 1.2 µm wavelength.

PRINCIPLES OF FIBER OPTICS

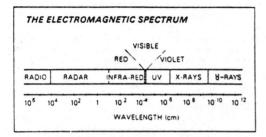

THE ELECTROMAGNETIC SPECTRUM

All glasses scatter light due to frozen in thermal fluctuations of constituent atoms. These cause density and hence refractive index variations within the material. It is believed that this intrinsic *Rayleigh scattering* represents the fundamental minimum limit to fiber attenuation. Since Rayleigh Scattering is found to vary inversely as the 4th power of wavelength (i.e. λ^{-4}) it follows that lower fiber attenuation occurs at longer wavelengths, and ideally light sources should be selected accordingly, consistent with adequate detector response.

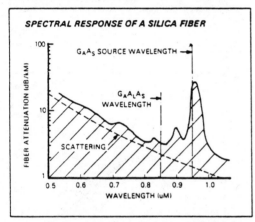

SPECTRAL RESPONSE OF A SILICA FIBER

In practice, of course, other considerations often assume greater importance. For example, it is particularly important to avoid certain attenuation peaks which occur with some types of fiber. One particularly severe peak occurs close to the GaAs source wavelength. This is caused by absorption in the core by hydroxyl ions and consequently is frequently referred to as a *water peak*.

As well as the *nominal* operating wavelength, the *spectral linewidth* (range of emitted wavelengths) of the source is important since each wavelength propagates at a different velocity. As will be seen, this limits the bandwidth of high data rate systems, and consequently semiconductor lasers are preferable to light emitting diodes as sources in such cases, since these exhibit narrower linewidths.

OPTICAL PULSE DISPERSION

Fibre optics communication is concerned with the transmission of information by varying or modulating the intensity of the light source at appropriate intervals. Fibers are most useful if they can faithfully support a rapid modulation frequency, or high data rate.

Apart from electronic limitations in the transmitter and

receiver, a restriction is imposed on the maximum frequency by the fiber itself.

Light is transmitted along a fiber by a multitude of different paths, ranging from one which is parallel to the axis to those propagating at angles close to the critical angle, with many in between. Each path at a different angle is termed a *transmission mode*.

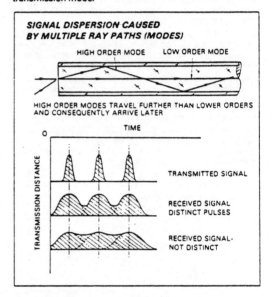

SIGNAL DISPERSION CAUSED BY MULTIPLE RAY PATHS (MODES)

It is clear that the distances travelled by various modes, and hence the times taken, are not equal.

Consequently, a short pulse of light, launched simultaneously into many modes, will have various transmission delays, and will arrive at the exit dispersed over an extended period of time.

This limits the maximum data rate, since a rapid train of pulses will merge into one another and may not be distinguishable. There will also be a reduction in amplitude caused by the pulse spreading. This is referred to as *intermodal dispersion*.

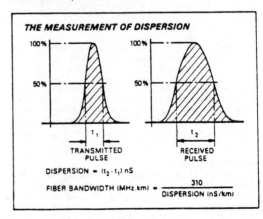

THE MEASUREMENT OF DISPERSION

DISPERSION = $(t_2 - t_1)$ nS

FIBER BANDWIDTH (MHz.km) = $\dfrac{310}{\text{DISPERSION (nS/km)}}$

PRINCIPLES OF FIBER OPTICS

The usual measure of dispersion is the increase in width of a Gaussian shaped pulse, measured at the half amplitude level. This is dependent on both the length and the NA of the fiber under test. Values are often quoted in ns/km.

Sometimes a fiber bandwidth is specified, corresponding to the upper frequency at which dispersion reduces the signal modulation amplitude by 3 dB. Typical values for a 0.5NA fiber are 50 ns/km or 7.75 MHz.km.

MODE CONVERSION

In real fibers, bends and minor irregularities at the core-cladding interface lead to conversion between high and low order modes, i.e. changes in ray angle.

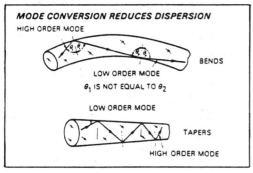

MODE CONVERSION REDUCES DISPERSION

Thus, to some extent, dispersion is reduced since all rays travel by high and low order paths along the fiber. This is particularly noticeable in longer lengths of fiber and in such cases dispersion may be assumed to increase in proportion to the square root of length rather than as a linear function.

For low frequency applications, dispersion is usually insignificant and maximum length will be limited by fiber attenuation. However, for long haul high bandwidth telecommunications, fiber dispersion may be predominant, and different fiber constructions have been evolved to ease the problem.

STEP AND GRADED INDEX FIBRES

The fibers considered so far exhibit a distinct change in refractive index between the core and cladding, and are called *step index fibers* for this reason.

Imagine a fiber constructed with many such changes in refractive index, with larger values towards the centre. A ray crosses the axis, approaching the circumference . . .

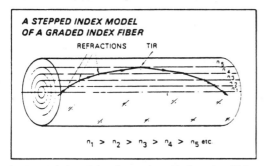

A STEPPED INDEX MODEL OF A GRADED INDEX FIBER

$$n_1 > n_2 > n_3 > n_4 > n_5 \text{ etc.}$$

At the majority of interfaces, the critical angle will be exceeded so that a *refraction* occurs. This, however, reduces

the grazing angle, and eventually conditions will be met for total internal reflection to take place. From then on, the ray will be refracted back towards the axis in a symmetrical manner.

If this model is replaced by a fiber in which the refractive index is continuously graded from a maximum at the centre to a minimum at the circumference, the previously stepped path is replaced by a smooth curve. Moreover, since the same phenomenon occurs on the opposite side of the axis, light is transmitted down the fiber following a smooth oscillating path.

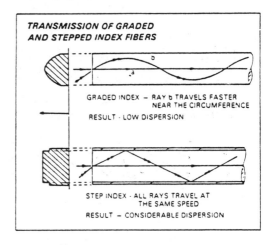

TRANSMISSION OF GRADED AND STEPPED INDEX FIBERS

There is still a path length difference between modes (a) and (b), but from the original definition of refractive index it is clear that ray (b) accelerates as it moves into the lower index medium near the circumference and generally travels faster than ray (a) in the high index centre. If the refractive index profile varies in a nearly parabolic manner, the increased speed of ray (b) exactly compensates for the greater path length, and the two rays, and consequently all rays, pass through the axis at the same instant.

It is difficult to fabricate fibers with the exact refractive index profile required, but a sufficiently close approximation can be made to give a useful factor of 10 or more reduction in dispersion. These fibers are called *graded index fibers*.

NUMERICAL APERTURE (GRADED INDEX FIBERS)

The numerical aperture of a step index fiber was defined earlier as:

$$NA = \sqrt{N_1^2 - N_2^2}$$

where N_1 = core index
N_2 = cladding index
($N_3 = 1$)

Since there is only a very small difference in refractive index between the core and cladding (typically 1 – 1.5%) very little error is introduced by expressing the NA in terms of this difference Δ and the average material index N using the approximation

$$NA = N\sqrt{2\Delta}$$

where $\Delta = \dfrac{N_1 - N_2}{N}$

In practice N may be taken as being equal to either N_1 or N_2, or the index of the raw material (say 1.4 – 1.5), whichever is

PRINCIPLES OF FIBER OPTICS

the most convenient. Thus, for example, a fiber with an index difference of 1% exhibits an NA of about 0.2.

A graded index fiber with a maximum index difference Δ exhibits an identical NA $= N\sqrt{2\Delta}$ on its axis, but lower values towards the periphery in line with the reducing refractive index. For a parabolic index profile the effective NA for the total core can be shown to be

$$NA_{eff} = N\sqrt{\Delta}$$

Thus it is clear that the total power acceptance of a graded index fiber (proportional to NA_{eff}^2) is half that of an equivalent step index fiber.

LIMITATION OF RAY THEORY—MONOMODE FIBERS

Although the concept of a ray travelling in a straight line provides a useful and quite accurate basis for describing the transmission properties of light on a large scale, a more detailed understanding can only be achieved by considering it in terms of a wave motion.

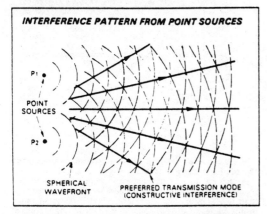

INTERFERENCE PATTERN FROM POINT SOURCES

P1

POINT SOURCES

P2

SPHERICAL WAVEFRONT

PREFERRED TRANSMISSION MODE (CONSTRUCTIVE INTERFERENCE)

We must recognise, for example, that the direct and reflected components of a wavefront interfere with one another, either constructively or destructively according to their relative phases. This defines a number of preferred ray propagation paths—transmission modes—which occur at discrete angles to the axis bounded by the numerical aperture of the fiber.

In other words, just like any other waveguide, the fiber is only capable of supporting a limited number of modes, this number being related to the fiber NA, its core diameter d, and the wavelength of light λ by the expressions

$$N \approx 0.5 \left(\frac{\pi.d.NA}{\lambda}\right)^2 \quad \text{— step index fibers}$$

$$N \approx 0.25 \left(\frac{\pi.d.NA^2}{\lambda}\right) \quad \text{— graded index fibers}$$

As long as the core diameter exceeds about 10 wavelengths, the number of modes is considerable, and their distribution is virtually indistinguishable from a continuum. Such fibers are termed *multimode*.

A significant feature of the transmission modes in the simplified model illustrated is that each successively higher order mode contains some light which has been delayed by one additional complete wavelength, thus introducing signal dispersion within that mode and contributing to the dispersion of the total wavefront. Clearly reducing the number of permissible modes will result in lower dispersion and hence higher bandwidth capability.

This principle is equally valid in optical fibers. A reduction in NA is possible, but there is a practical limit to this, beyond which the light is insufficiently well guided and escapes at bends of even quite moderate curvature. However, reducing the core diameter also leads to a reduction in the number of modes, and in particular only one can propagate if the diameter is less than the critical value

$$d_c = \frac{2.4\lambda}{\pi NA}$$

This remaining mode is labelled HE_{11} and is significant in not being subject to intermodal dispersion. Fibers made to this criterion are called *single* or *monomode* and are capable of supporting extremely high bandwidths (potentially 50 GHz.km). Bandwidth is ultimately limited by a second form of dispersion known as *intramodal*. This is the result of refractive index (and hence propagation velocity) variation with light source wavelength.

Hence for the full exploitation of monomode fibers, light sources are required which exhibit very narrow spectral linewidths. They must also be very small and bright in order to permit efficient coupling into the very small core diameters of these fibers (typically 3-5 microns).

4 FIBER CONSTRUCTIONS

Optical fibers are commonly manufactured from either plastic, silica, or glasses containing lead, sodium or boron compounds.

PLASTIC FIBERS

Fibers constructed from transparent plastic offer the advantages of large diameter (as large as 1 mm), flexibility and ease of termination (ends can be prepared by cutting with a hot razor blade). However, the high intrinsic loss of the materials used limits them to applications involving transmission distances of only a few metres, and to environments protected from extremes of temperature.

GLASS FIBERS

Glass is the traditional medium used in the manufacture of optical fiber, relying until recently on materials used for high quality lenses.

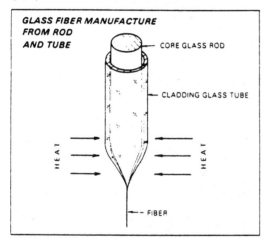

GLASS FIBER MANUFACTURE FROM ROD AND TUBE

CORE GLASS ROD

CLADDING GLASS TUBE

HEAT

HEAT

FIBER

For many years fibers have been manufactured by the 'rod and tube' method, in which concentric billets of core and cladding glasses are heated in a furnace and drawn out into a thin fiber. These fibers used to exhibit losses in excess of 500 dB/km, but research into materials and processes has reduced this to levels compatible with short and medium range communication requirements.

Alternatively, glass fibers can be extruded directly from

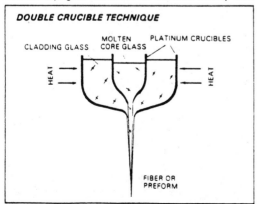

DOUBLE CRUCIBLE TECHNIQUE

CLADDING GLASS

MOLTEN CORE GLASS

PLATINUM CRUCIBLES

HEAT

HEAT

FIBER OR PREFORM

concentric platinum crucibles containing the core and cladding glasses in a molten state. This is capable of producing very low loss fibers and has the potential advantage of being a continuous process, there being no limit imposed on the length of fiber produced if the raw materials are replenished at regular intervals.

SILICA FIBRES

Silica (SiO_2) is a material which occurs naturally in the form of quartz, or can be produced synthetically, and which exhibits very low intrinsic optical loss making it an ideal candidate for the manufacture of fibers. However, it also has a very low refractive index so that the requirement for a still lower index cladding has necessitated the development of special manufacturing techniques.

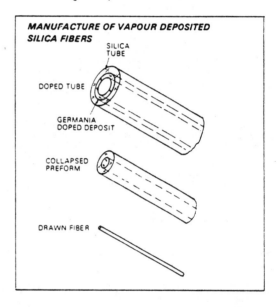

MANUFACTURE OF VAPOUR DEPOSITED SILICA FIBERS

SILICA TUBE

DOPED TUBE

GERMANIA DOPED DEPOSIT

COLLAPSED PREFORM

DRAWN FIBER

One of these techniques, vapour deposited silica (VDS), involves laying radial layers of germania-doped silica on the inside of a silica tube to form a region of increased refractive index. This forms the core when the tube is later collapsed into a solid rod and subsequently drawn into a fiber.

These layers are produced by passing suitable gases through the tube whilst being heated. By varying the constituents of these gases, both core and cladding regions can be formed. Alternatively, the refractive index profile may be changed gradually, to produce graded index fibers.

Not surprisingly, this technique is fairly expensive, but produces very low loss fibers (< 1 dB/km), 5 dB/km probably representing a realistic commercial value at the present time.

PLASTIC CLAD SILICA (PCS) FIBERS

A fundamentally different solution to the problem of providing a lower index cladding is achieved by applying a thin layer of low refractive index silicone resin to a pure silica core after it has been drawn into a fiber.

This is an attractive technique, producing large core diameter fibers at costs determined principally by the raw materials used. Losses are not among the lowest obtainable due to absorption within the lossy cladding—some ray penetration

FIBER CONSTRUCTIONS

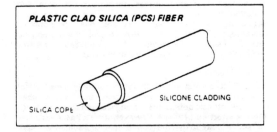

PLASTIC CLAD SILICA (PCS) FIBER

SILICA CORE

SILICONE CLADDING

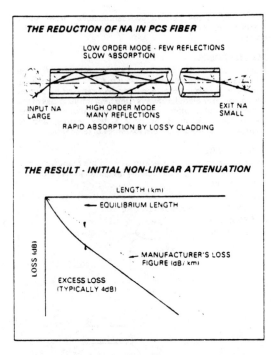

THE REDUCTION OF NA IN PCS FIBER

LOW ORDER MODE - FEW REFLECTIONS
SLOW ABSORPTION

INPUT NA
LARGE

HIGH ORDER MODE
MANY REFLECTIONS

EXIT NA
SMALL

RAPID ABSORPTION BY LOSSY CLADDING

THE RESULT - INITIAL NON-LINEAR ATTENUATION

LENGTH (km)

EQUILIBRIUM LENGTH

MANUFACTURER'S LOSS
FIGURE (dB/km)

LOSS (dB)

EXCESS LOSS
(TYPICALLY 4dB)

of the cladding always occurs during total internal reflection—but are acceptable for many applications.

This absorption in the cladding leads to a unique phenomenon. Rays travelling at high angles to the optical axis (high order modes) clearly intersect the interface more frequently than lower orders, and hence are absorbed to a greater extent over a given length. This leads to a gradual reduction of the effective numerical aperture of the fiber over the first few tens of metres, until an equilibrium NA is reached when the loss of high order modes is balanced by mode conversion from lower orders.

The result of this NA change is a non-linear attenuation characteristic as far as the equilibrium length. Consequently, estimates of path loss with this fiber must include an excess loss calculated from the squared ratio of the initial to equilibrium NAs. Although such estimates will be pessimistic on short lengths, it is inadvisable to use a lower figure as the exact shape of the non-linear region is indeterminate, being significantly influenced by the degree of curvature present in the fiber.

A reduction in NA is also seen at low temperatures, because, as the silicone resin is cooled, it becomes denser and its refractive index increases. Thus the environmental performance of PCS is limited, but it is nevertheless an ideal choice for many industrial applications.

5   FIBER CABLES

PRIMARY AND SECONDARY COATINGS

The majority of fibers exhibit considerable tensile strength, which for *freshly* drawn silica fibers can exceed that of best quality steels. Unfortunately, surface damage caused by handling or even atmospheric attack rapidly leads to a dramatic degradation in strength. This can be prevented by protecting the fiber with a thin plastic coating, immediately after manufacture, and usually as an on-line process. This coating is referred to as a *primary coating*. Since the silicone resin used in the manufacture of PCS fiber makes an excellent primary coating, this is frequently used with other fibres.

Although the primary coating maintains the intrinsic tensile strength of the fiber, being relatively thin and soft it does little to protect the fiber from external mechanical damage. Consequently a further *secondary* coating is added to give protection and improve handling quality.

Two schools of thought have arisen regarding the form that the secondary coating should take—namely whether or not it should be a loose or a tight fit over the primary coated fiber. This is because the addition of protective coatings must not be detrimental to the optical properties of the fiber, and in particular must avoid introducing *microbending loss*, i.e. additional attenuation caused by local inhomogeneities in the coating material which lead to a distortion of the fiber.

Loose tube constructions provide a simple solution to this problem, and accordingly have been adopted by several fiber manufacturers. Unfortunately this design suffers from other problems due to the instability of the plastic coating material. For example, when this shrinks with age or changing temperature (as invariably happens as longitudinal extrusion stresses are relieved) the fiber is forced to take the form of a helix within the loose tube. In severe cases this also can cause increased attenuation because of the minute but continuous leakage of light from the curved fiber. The process is further complicated by the high coefficients of expansion of the plastics used.

ITT have adopted the alternative tight tube construction avoiding the problems of microbending by strict material and process control. Although shrinkage of the coating material can still occur, this merely puts the fiber into compression and has no detrimental effect on attenuation or strength. It does, however, produce an additional problem at each end of the fiber, since the shrinkage causes the fiber end surface to protrude beyond the end of the protective coating. This phenomenon has been termed *'growing-out'* and puts the fiber at risk in connectors. Consequently, termination techniques have had to be developed which are sufficiently sound to resist this movement and the forces involved. This problem overcome, the result is a very stable structure.

CABLE CONSTRUCTION

The overall design of a ruggedised fiber cable depends on the application and the number of communication channels required, but invariably features some form of tensile strength member and a tough outer sheath to provide the necessary mechanical and environmental protection.

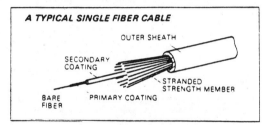

A TYPICAL SINGLE FIBER CABLE

OUTER SHEATH

SECONDARY COATING

STRANDED STRENGTH MEMBER

BARE FIBER

PRIMARY COATING

Ruggedised single fiber cables usually employ a fibrous external strength member (such as Kevlar) which is either laid helically, or braided around the secondary coated fiber. This is surrounded by a plastic outer jacket for additional crush and abrasion protection.

Multifiber cables are made in a variety of configurations. The simplest involves grouping a number of ruggedised single fiber cables within a further outer jacket. This is not a low cost construction, but it is strong and has the advantage that, when the outer jacket is removed at each end of the cable the individual ruggedised cables can be terminated in separate single way connectors. Thus a very high degree of immunity to mechanical and handling damage can be achieved in the region which is often subjected to considerable stress.

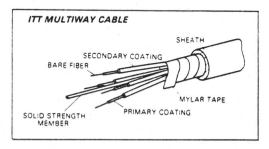

ITT MULTIWAY CABLE

SHEATH

SECONDARY COATING

BARE FIBER

SOLID STRENGTH MEMBER

MYLAR TAPE

PRIMARY COATING

ITT multifiber cables normally feature a solid central strength member with secondary coated fibers laid in a helix around it. The strength member can be steel in applications where the cable is not required to have isolating properties. Cables to this design are available with either 4 or 8 communication fibers, or featuring a mix of optical and electrical conductors.

Alternatively, an external fibrous strength member can be used, or, for highest tensile load requirements (e.g. aerial cables) a combination of both.

When the outer jacket of a multifibre cable is removed, the less well protected secondary coated fibers will usually be exposed. In order to minimise the risk of mechanical and environmental damage, cables should ideally be terminated with multiway connectors featuring appropriate cable and strength member tie-off facilities in order to provide protection at this point.

CABLE TESTING

It is an interesting feature of the progress made in the development of fiber optics cables that the specifications now being achieved exceed those of many conventional communication cables.

a) Tensile Strength

The tensile strength of freshly drawn glass fibers is comparable to that of any high tensile material including steel, and the introduction of primary coatings has virtually eliminated degradation caused by mechanical and atmospheric damage.

The failure of fibers now being manufactured is generally caused by microscopic surface flaws which occur infrequently and at random intervals. Consequently it is essential that sufficient fiber is tested (i.e. a large number of short samples or a smaller number of long samples) in order to obtain meaningful values of ultimate tensile strength.

In order to illustrate the statistical failure mechanism within the fiber, the results of many identical tensile tests are plotted as a Wiebull chart, i.e. accumulated number of failures versus breaking strain (strain is used as the load parameter since this

FIBER CABLES

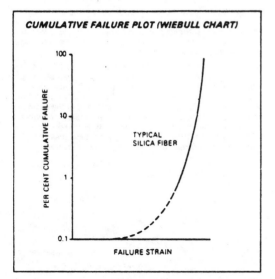

CUMULATIVE FAILURE PLOT (WIEBULL CHART)

TYPICAL SILICA FIBER

PER CENT CUMULATIVE FAILURE

FAILURE STRAIN

is independent of fiber diameter). On this chart, an ideal fiber featuring a constant breaking load is represented by a vertical line. Less than ideal fibers will show variable degrees of slope.

Many commercial silica fibers will withstand proof strains exceeding 1% with an ultimate failure strain of greater than 5%. Consequently, cables are designed with sufficient tensile stiffness so that, at the maximum rated load, the fiber strain is limited to the safe lower value of 1%. Typical ITT multiway cables are rated at 1500 N (150 kg) and rugged single cables at 500N.

b) Crush Resistance

Resistance to crushing loads is provided primarily by the cushioning effect of the outer cable sheath, although even secondary coated fibers exhibit considerable tolerance to physical abuse. The cables are subjected to a gradually applied load simulating the pressure that could be generated by the heel of a 100 kg man. A second test involves a radiused anvil representing obstructions such as a ladder rung. Both multiway and rugged single fiber cables have been tested in this manner without failure after 1000 cycles.

c) Impact resistance

Repeated impact can induce failure by causing permanent changes in the plastic components of a cable. In an ITT test based on a U.S. Military specification, blows are delivered from a height of 100 mm at a rate of one per second by a 12.5 mm radiussed head with the cable on a flat steel anvil. Load can be varied from 500 g to 4 kg.

Despite severe cable distortion, all fibers in the multiway cable construction have survived 1000 blows at 2 kg. It is of interest to note that in cables containing a mix of optical fibers and copper conductors, the copper elements all failed before the fibers. In designs, with steel strength members, the fibers have outlasted the steel.

On a more mundane, yet possibly more convincing level, an eight fiber multiway cable was laid across a busy car park entrance. Some 40,000 vehicles and several sceptical boots later, the cable was intact, with a deep groove worn in the tarmac surface.

d) Resistance to Bending Failure

A cable may be subjected to a considerable degree of reeling

during its life and therefore the effects of bending must be considered during its design. Conventional cables usually specify a minimum bend radius of 20 cable diameters but fiber cables manufactured by ITT are subjected to a more rigorous bend test in which four turns of cable are wound in a grooved mandrel with a radius of approximately 5 D at a tension of 100 N. Continuous winding and unwinding has shown no failure up to 100 cycles.

e) Environmental Tests

Tests involving temperature and humidity cycling have been designed to suit various applications. Current cable design show very little attenuation change over −15°C to +60°C and are unaffected by humidity.

The primary coated fiber is stable over a very wide temperature range (−50°C to 200°C) and the overall performance of a cable is governed by the choice of plastic sheathing materials. High temperature cables require the use of expensive fluoropolymers and consequently are reserved for essential applications.

STATIC FATIGUE

Unlike many conventional engineering materials which exhibit fatigue failures when subjected to cyclic strains, glass and silica fibers will eventually fail if a *constant* tensile strain less than the normal failure strain is present for a sufficiently long period of time. This is called *static fatigue* and is caused by the propagation of cracks from surface flaws as a result of a process known as *stress corrosion*.

Research into this phenomenon has indicated that fatigue life (i.e. the mean life to failure) is a very sensitive function of tensile strain – typically (strain)$^{-14}$ – although, not surprisingly there is some difference of opinion regarding the exact relationship.

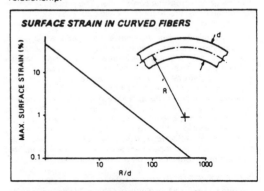

SURFACE STRAIN IN CURVED FIBERS

MAX. SURFACE STRAIN (%)

R/d

Long term strains are most likely to occur in fiber bends, and here a simple relationship connects maximum tensile strain to bend radius R. Hence

$$\text{Fatigue Life} \propto \left(\frac{R}{d}\right)^{14} \qquad \text{where d = fiber diameter}$$

The sensitivity of this function is dramatically illustrated by considering the effect of halving an appropriate bend radius. Life expectancy is then reduced by a factor 2^{14} (say from 25 years to 13 hours. Halved again, to 3 seconds!)

Static fatigue failures are unlikely to be experienced by the user since cables are designed to restrict minimum bend radius to a safe value. Tests have also shown that cyclic bending is not a significant contributory factor. Static fatigue is of most significance in the design of right angle connectors where the constraints of size dictate small bend radii.

6 FIBER TERMINATION AND CONNECTORS

FRESNEL LOSS

To avoid unnecessary loss of light, the input and exit surfaces of an optical fiber must be smooth and perpendicular to the axis.

Even with optically flat ends, a small proportion of the incident light is reflected back, this being an unavoidable phenomenon associated with a step change in refractive index called *Fresnel reflection*. The magnitude of this reflection is about 4% for a typical glass-air interface, and is the same whether light is entering or leaving the fiber. Consequently, a fibre-to-fibre coupling, involving two interfaces, introduces a total loss of nearly 8%, or about 0.35 dB.

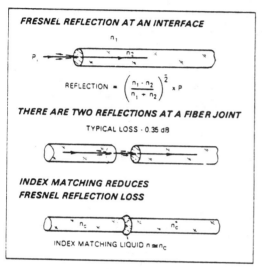

FRESNEL REFLECTION AT AN INTERFACE

$$\text{REFLECTION} = \left(\frac{n_1 - n_2}{n_1 + n_2}\right)^2 \times P$$

THERE ARE TWO REFLECTIONS AT A FIBER JOINT

TYPICAL LOSS - 0.35 dB

INDEX MATCHING REDUCES FRESNEL REFLECTION LOSS

INDEX MATCHING LIQUID $n \simeq n_c$

Fresnel reflection can be reduced to a very low level by filling the gap between adjacent fiber ends with an *index matching* liquid, that is, a liquid which exhibits the same refractive index as the fiber core.

Index matching is not considered to be a practical approach in demountable connectors since dirt collection is a major problem with such liquids. However it can be successfully employed in more permanent joints such as a mechanical splice.

FIBER TERMINATION

A bare fibre end is quite fragile and, if not adequately protected, prone to damage in service. Consequently it is usually terminated in a suitable receptacle called a *ferrule*. This must provide accurate location of the fiber on its geometric axis, and be sufficiently well secured to both fiber and secondary coating to resist the axial 'growing-out' movement referred to earlier.

To prevent growing-out, the fiber is sometimes cemented into the ferrule with an epoxy resin, and subsequently ground flat and polished to an optical finish. This process, referred to as *cement and polish*, has been used by ITT with a wide variety of fiber types, and its reliability has been verified over a range of environmental conditions. Its disadvantage is that a delay is introduced whilst the epoxy resin cures, but this has been reduced to an acceptably short period (5-10 minutes) by heating the assembly in a special fixture, so that a completed termination can be achieved in about 20 minutes.

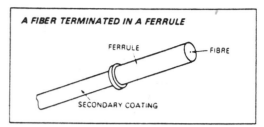

A FIBER TERMINATED IN A FERRULE

FERRULE · FIBRE

SECONDARY COATING

An alternative termination technique has also been developed by ITT for use with large core PCS fibers. With such fibres, the cement and polish process is not suitable, since external adhesion to the silicone resin cladding is not good enough to resist growing-out forces. Neither can the cladding be removed to permit direct adhesion to the fiber core, since the higher index adhesive would prevent the vital total internal reflection in that region, and lead to considerable loss of light.

For the termination of PCS fibre, the fiber end is prepared by *cleaving*. This is a technique which has evolved from the discovery that a fiber, scribed with a diamond blade whilst subjected to a particular combination of tensile and bending stresses, will break in a controlled manner, usually leaving a flat mirror finished end. Various hand tools have been developed to exploit this principle.

After preparation, the fiber is located inside a special ferrule, with the end butting against a very thin optical 'window' which is fixed to the front of the ferrule. This resists growing out. In addition, the ferrule is secured to the secondary coating by crimping, and an index matching 'gel' is used to eliminate the additional Fresnel reflections which would otherwise occur at the fiber-to-window surfaces.

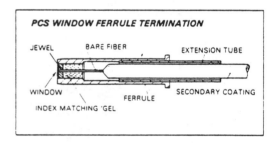

PCS WINDOW FERRULE TERMINATION

JEWEL · BARE FIBER · EXTENSION TUBE

WINDOW · FERRULE · SECONDARY COATING

INDEX MATCHING 'GEL'

The window termination technique is easy and quick to perform, and is particularly suitable as a field technique. Currently its use is limited to large core PCS fibers (i.e. greater than 200 microns) since the effective gap between the fiber ends which is introduced by the windows causes unacceptably high loss between other small core fibers.

BUTT-JOINT CONNECTORS

Coupling between fibers is most frequently accomplished by a direct butt joint between prepared fiber ends. To minimise loss of light at the joint *(insertion loss)*, the two fibers must be accurately aligned, both axially and in an angular sense, and with the minimum possible gap between the ends, but avoiding actual contact to prevent damage from surface abrasion.

The most difficult parameter to control is the axial alignment since, with typical core diameters often less than 100 microns, even quite small misalignments will result in appreciable insertion loss.

FIBER TERMINATION AND CONNECTORS

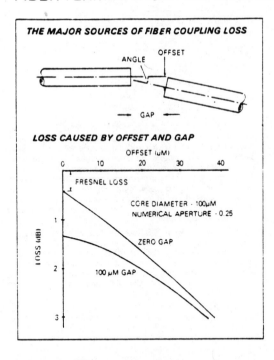

THE MAJOR SOURCES OF FIBER COUPLING LOSS

ANGLE OFFSET

GAP

LOSS CAUSED BY OFFSET AND GAP

OFFSET (uM)

FRESNEL LOSS

CORE DIAMETER · 100μM
NUMERICAL APERTURE · 0.25

ZERO GAP

100 μM GAP

LOSS (dB)

**SOURCES OF FIBER MISALIGNMENT
(FERRULE-SLEEVE COUPLING)**

FERRULE

FERRULE-SLEEVE
CLEARANCE

FERRULE I/D - O/D
ECCENTRICITY

FIBER - FERRULE
CLEARANCE

CORE - CLADDING
ECCENTRICITY

ALIGNMENT SLEEVE

RESULTING COUPLING LOSS DISTRIBUTION

FIBER · 82 μM CORE · 0.5 NA
CONNECTOR · OCN101

90% CONFIDENCE LEVEL

FREQUENCY

0.5 1.0 1.5 2.0 2.5 3.0
FRESNEL
LOSS   COUPLING LOSS (dB)

The lowest insertion losses can be achieved by direct alignment of bare fibers (in a V-groove, for example) but the design of a successful connector using this principle is difficult because of the need to protect the exposed fiber in the uncoupled state. More frequently the fiber is totally enclosed in a ferrule with only the flat front surface exposed. Two similar ferrules are aligned by locating in, for example, a closely fitting sleeve. This principle is employed in ITT connectors.

The success of this technique is limited by the manufacturing tolerance that can be achieved in the various components. For example, fiber-to-ferrule and ferrule-to-alignment sleeve clearances must necessarily exist, both of which contribute to the total fiber misalignment. Similarly, the ferrule itself must be allowed a manufacturing tolerance on concentricity. Each of these effects exists in both halves of a mated connector, and consequently there are at least six independent sources of fiber misalignment.

The maximum possible misalignment results from the arithmetic sum of all the contributions. In practice, however, each exhibits a variable magnitude and a random direction relative to its neighbours, and a realistic assessment takes this into account. The result is a statistical distribution of connector loss. In the example shown, although a worst case value of 3 dB is *possible*, a 2 dB maximum can safely be assumed at a high statistical confidence level (> 95%) and appreciably lower values with only a slightly reduced level of confidence. Thus, in practice, acceptable performance can be achieved with even the smallest fiber sizes.

EXPANDED BEAM CONNECTORS

An alternative to direct butt joints between fibers is offered by the principle of expanded beam. If the fiber is placed at the focus of a converging lens, an essentially collimated (parallel) beam is produced, the degree of collimation depending on the relative sizes of the lens and fiber.

A second similar lens, placed in line some distance away, will refocus the beam to a spot at which the receiving fiber is located. This forms an expanded beam joint.

The advantages of this technique accrue from the relatively large diameter and good collimation of the beam between lenses, making optical connectors using this principle

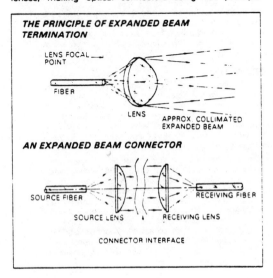

**THE PRINCIPLE OF EXPANDED BEAM
TERMINATION**

LENS FOCAL
POINT

FIBER

LENS APPROX. COLLIMATED
EXPANDED BEAM

AN EXPANDED BEAM CONNECTOR

SOURCE FIBER   RECEIVING FIBER

SOURCE LENS RECEIVING LENS

CONNECTOR INTERFACE

FIBER TERMINATION AND CONNECTORS

correspondingly *less* sensitive to axial misalignment and gap. In addition, the larger lens surface is less likely to be obscured by small particles of dirt, and is generally easier to clean.

The trade-off for these benefits is that expanded beam connectors are proportionally *more* sensitive to small *angular* misalignments. Nevertheless expanded beam will probably become an important feature of future optical communication links, particularly those subjected to severe operating environments.

CLADDING MODES

The silicone resin which constitutes the optical cladding of a plastic clad silica fiber is commonly used as a primary coating in other fiber constructions featuring conventional core and cladding regions. With these fibers it is frequently observed that light is guided not only within the core but also by the cladding. This is the result of total internal reflection which occurs at the cladding-primary coating boundary.

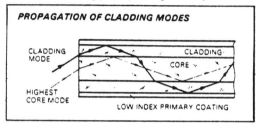

PROPAGATION OF CLADDING MODES

Whilst such *cladding modes* might be considered to be a bonus in terms of transmitted power, in practice they are quite lossy and decay to negligable proportions in lengths exceeding a few hundred metres. They are, however, present to a significant extent in short lengths of fiber, and this can lead to misleadingly optimistic results in connector evaluation trials (since the loss caused by a given fiber misalignment is correspondingly smaller if the entire core plus cladding area is illuminated).

Similarly, measurements of transmitter power present in short lengths of the fiber will be artificially high if cladding modes are not removed.

For example, a typical telecommunications fiber has a 50 μm core diameter and 125μm overall cladding diameter. Assuming that the core NA is 0.25, and that of the cladding is the same as for a PCS fiber (i.e. 0.39) the ratio of useful core power to the total fiber power is given by:

$$\left(\frac{50}{125} \times \frac{0.25}{0.39}\right)^2 = 6.6\% \text{ i.e.} - 11.8 \, dB$$

Cladding modes can be stripped by ensuring that a sufficiently long length of fiber is bared of its primary coating and surrounded by a high refractive index medium. This can be conveniently accomplished in the ferrule termination using normal epoxy resin.

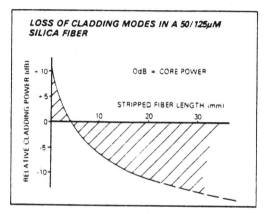

LOSS OF CLADDING MODES IN A 50/125μM SILICA FIBER

Although the removal of cladding modes is an exponential process (since the lower order modes make contact with the boundary very infrequently) it has been found that a stripped length of 20–30 mm will eliminate sufficient cladding power for normal measurement purposes.

Cladding mode removal should not be regarded as an essential requirement of an installed link. It may, however, be necessary during evaluation.

Chapter 6

LASER RADIATION
AVOID DIRECT
EYE EXPOSURE

HeNe 5 mW max

CLASS IIIA LASER PRODUCT

Universal Helium-Neon Laser Power Supply (HNE3)

T HIS PROJECT SHOWS HOW TO CONSTRUCT A VERY VERSATILE AND USEFUL power supply for powering most He-Ne gas discharge tubes from 0.1 to 10 mW. A special trigger circuit is included for "hard to start" tubes. Internal ballast resistors are noted for the particular port or jack on the rear of the device. It may be necessary to use other ballast resistors for odd parameter tubes. A meter is shown for measuring tube current and can be used for final tailoring of the ballast resistors. Housing is shown in an attractive painted metal cabinet but may be changed per the builder's needs.

SPECIAL NOTE BEFORE ASSEMBLING

The unit is shown constructed with a key switch, emission indicator, time delay and remote control port for BRH compliance. Interlocks and labeling must be taken into consideration when powering a tube of a particular class. If in question you may contact the Bureau of Radiological Health and Welfare Office in your area. Proper labeling is shown further into these instructions for the range of tubes capable of being powered.

Also take note of the surplus laser tubes available at reduced cost, as these are usually second rate being hard to start, have poor beam quality, reduced output, and are aged beyond their useful life. It is suggested that you purchase a new tube with a spec sheet and all guarantees or proper performance may never be realized by the inexperienced laser experimenter.

CIRCUIT DESCRIPTION

Key switch S1 switches 115 Vac to the primary of transformer T1. Indicator lamp

(NE1) and current-limiting resistor (R1) shows when the system is energized. A three-wire cord set (CO1) effectively grounds the case (CA1) of the unit. The 115 Vac is stepped up to 1200 Vac by T1 and is rectified and voltage doubled to 2500 Vdc by diodes (D1 through D8) and capacitors (C1 and C2). Resistors (R2 through R9) are necessary to provide even distribution of the reverse voltage across the diodes.

A high-voltage trigger pulse of 12 Vdc is provided by the action of pulse transformer (T2). This transformer is pulsed in its primary via a capacitor discharge circuit consisting of a capacitor (C6) and a silicon-controlled rectifier (SCR1). C6 is charged to approximately 250 kVdc by the voltage-divider section consisting of resistors (R10, 11, 12). Diode D9 provides a return path for the negative overshoot of the SCR when turning on. Note that the voltage at R10 is approximately 1250 as it is at the junction of C1 and C2. SCR1 is triggered by the breakover action of the diac (DI1). This unilateral device supplies the positive-going gate voltage pulse necessary for turning on SCR1. The positive voltage of *30* to *40* volts for breaking over DI1 is supplied by the charging action of capacitor (C5) and its charging resistor (R13). These components also supply the time delay necessary by the factor T = CV/I. Note that the current *I* is considered to be from a constant-current source due to the high feed voltage.

The voltage is clipped at 45 volts by zener diodes (Z1, 2, 3). A disabling circuit consisting of a transistor (Q1) is connected to shunt the trigger voltage built up across C5. Q1 is turned "on" when the laser tube turns "on" providing a forward bias by the voltage developed across resistor (R18). Resistor (R17) limits the base current to that necessary for positive turn-on of Q1. Resistors (R15, 16) limit the current in these respective circuits. A normal trigger function is provided by push-button switch (S1) and resistor (R14). This is sometimes necessary when testing older or hard to start tubes. The high voltage pulse output of T2 is rectified by a diode (D10) and a capacitor (C3). Note that the secondary of T2 along with D9 carries the laser tube current once the system is triggered on.

Output jacks J2, 3, 4 supply various amounts of current for certain power levels of tubes. These values are a function of their associated ballast resistors (R19 through R28) and should be tailored to the tube manufacturer's specs if continued use is required.

Remote control is provided by leads designated (WN1). These are normally connected together via a wire nut. Jack (J1) supplies no ballast and should be avoided unless a known external ballast is prewired into the laser tube enclosure. Jack (J5) is the negative or common returns of all the outputs.

CONSTRUCTION STEPS

1. Identify all parts in Table 6-1. Match electrical components to values shown Fig. 6-1.

2. Fabricate board for clearance of the mounting screws of SW1.

3. Layout board as shown in Fig. 6-1 through 6-3. Note the dashed lines indicating the connections of the components. Use the actual individual component leads where possible. Observe polarity of all semiconductors. Pay particular attention to the center pin of T2, D10, and C3. Maintain as much clearance as possible from other components as the voltage pulse when igniting the tube can be in excess of 10 kV. Note T2 is mounted with its output (center pin) facing upwards. Secure to perfboard via other leads and several dabs of liquid silicon, etc.

Table 6-1. Parts List for HNE3.

R1	1	100 kΩ 1/4-watt resistor
R2,3,4,5 6,7,8,9	8	5.6 MΩ 1/4-watt resistor
R10,11	2	1 MΩ 1-watt resistor
R12,13	2	510 kΩ 1-watt resistor
R14	1	1 kΩ watt 1/4-resistor
R15,16,17	3	100 ohm 1/4-watt resistor
R18	1	220 ohm 1-watt resistor
R19,20,21 22	4	20 kΩ-watt resistor
R23,24,25 26,27,28	6	120 kΩ 2-watt resistor
*C1,2	2	.25µF/4 kV (special) caps
*C3	1	.001µF/15 kV ceramic HV caps (special)
C5	1	10 µF/50 V elect
*C6	1	3.9µF/350 V pulse (special)
D1-8.9		1000 V 1 A rect diode 1N4007
*D10	1	20 kV HV rect HX100 (special)
Z1,2,3	3	15-volt zener 1N5245
DI1	1	25-volt diac
Q1	1	NPN2222 NPN transistor
NE1	1	Neon with leads
SCR1	1	C107D silicon-controlled rectifier
*T1	1	1200-volt/40 mA power transformer (special)
*T2	1	11 kV pulse transformer (special)
CO1	1	3-wire line cord
J1-5		Banana jacks 1/4″ hole
PB1		Perfboard 4-1/2″ × 4″
CA1		Case enclosure 9″ × 5-1/4″ × 2-3/4″
WN1	2	Small wire nuts
S1		Key switch (nonremovable when "on")
S2		Push-button switch
BU1		Neon lamp bushing
WR1		Wire
BU3,2	2	Strain bushing for power cord and "remote" wires
SW1/NU1	6	6-32 × 3/8 self tapping screws
SP1	3	3/8″ to 1/2″ plastic spacers—tap holes with screws
LAB1		Laser label Class III B
LAB2		HV label
LAB3,4		Laser labels as required should you use them (See section on labels and compliance requirements)
SW2	4	Nuts for attaching T1

Complete kit of above is available from Information Unlimited, P.O. Box 716, Amherst, NH 03031. Write or call (603) 673-4730. Parts marked with an asterisk are available individually.

LT1—Laser tube 0.2 to 5 mW. Please take heed when it comes to obtaining a laser tube. Be very cautious of the surplus market and "special" low cost deals often found in mail order catalogs. While these tubes may lase and produce coherent light they often are at reduced output, have very poor beam quality, are hard to start, and subject to fading out after a period of time. Also many are not certified and may present a compliance problem. It is strongly suggested to purchase a new laser tube from a reputable company complete with warrantee and instructions. This may cost more money but in the long run will provide you with the full benefit of an optimized system both in power and beam quality. Our first choice tubes are from Aerotech and we stock them at all times. It is suggested that you write or call us for present price and delivery information.

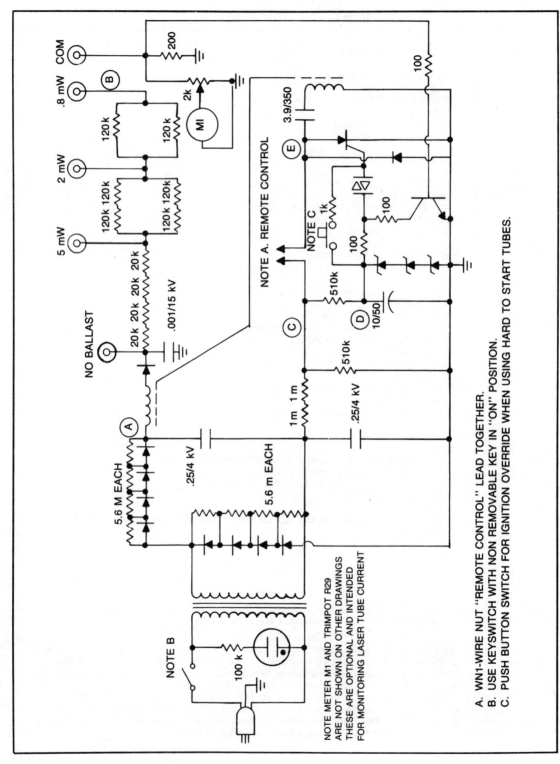

Fig. 6-1. HNE3 parts value universal hene laser.

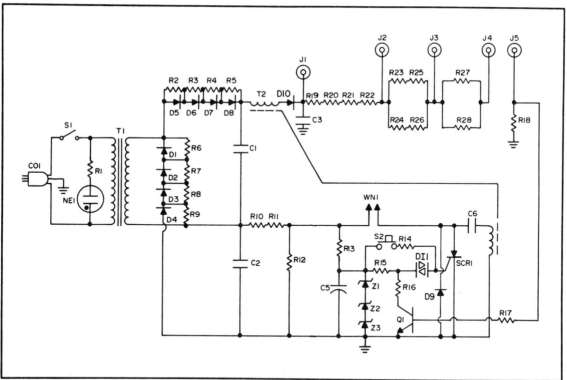

Fig. 6-2. HNE3 universal hene laser parts ID.

4. Attach the following leads to the board (see Fig. 6-4).

☐ Lead from R10 to C1, C2 (8″ #20 vinyl)
☐ leads referred to as "remote" (8″ #20 vinyl)
☐ Leads to S2 (6″ twisted pair of #24 vinyl)
☐ Leads to J5, J4, J3, J2, J1 (3″ #20 vinyl)
☐ Lead from T1 to C1 and C2 (6″ #20 vinyl)
☐ Leads to C1 and C2 (8″ #20 vinyl)

5. Fabricate front and rear panel of CA1 as shown Fig. 6-5. Trial position these components for proper fit and clearance. Layout is reasonable critical.

6. Fabricate bottom plate of CA1 for mounting of transformer (T1) and power board Fig. 6-3. Match holes carefully before drilling.

7. Assemble components as shown Fig. 6-6. Note plastic spacers (3) for mounting assembly board and the self-threading screws. Note transformer T1 has spacers included with it and only requires four nuts.

8. Finally wire as shown Fig. 6-4. Note wire nuts for connecting CO1 cord to transformer T1. Pay attention to wiring of NE1 and R1 as these points are at 115 Vac and must be thoroughly insulated from potentially shorting to metal case CA1.

9. Carefully check wiring for proper routing, clearance, and quality of solder joints. Unit is now ready to test. See Figs. 6-7 and 6-8.

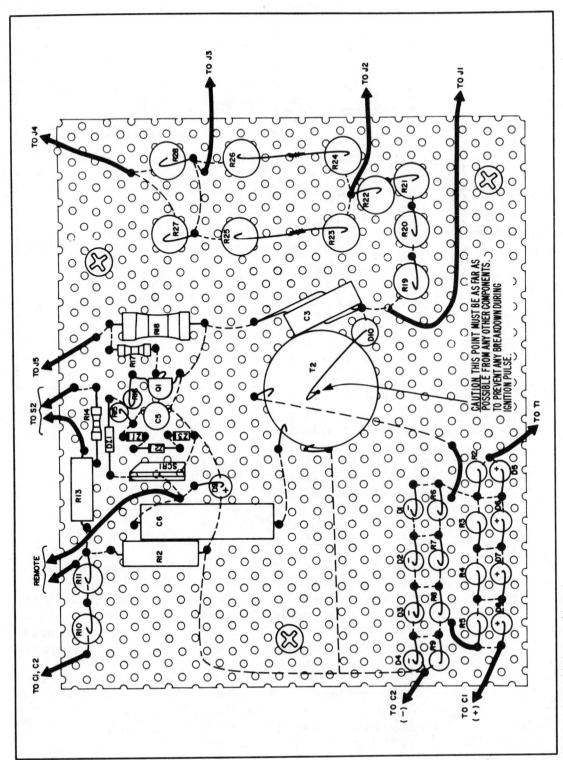

Fig. 6-3. Assembly layout of board.

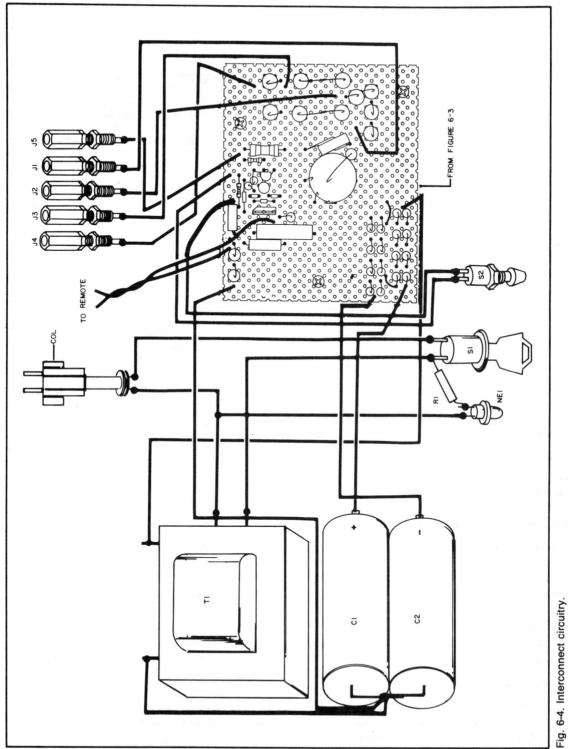

Fig. 6-4. Interconnect circuitry.

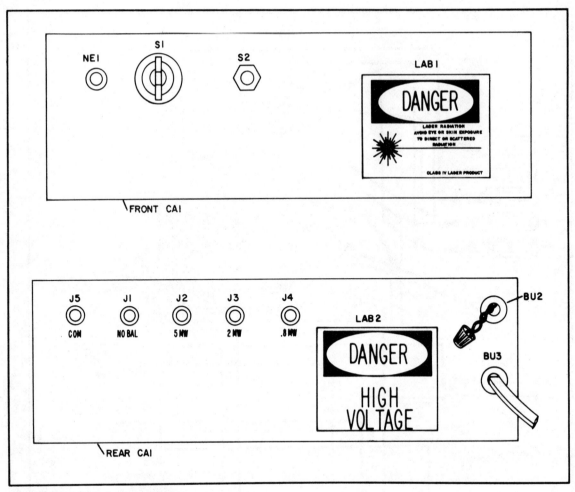

Fig. 6-5. Front and rear panel layout.

TESTING PROCEDURE AND SETUP

10. Connect a high-voltage meter to test point A Fig. 6-1 and measure 3700 volts when key switch S1 is energized. Also neon lamp should ignite indicating "power on."

11. Deenergize circuit and connect two series 150 kΩ2-watt resistors for 300 kΩ 4-watt in series with a millimeter between jack (COM) and jack (.8). Set to 10 mA range and measure approximately 6 mA. Also measure 3300 volts at test point A with this load current.

12. Measure approximately 1700 V at test point B, and 200 volts at test point C. Measure 30 volts at test point D with resistor R18 shorted. Note that this allows ignition to fire.

13. Connect voltmeter to test point E and note voltage switching between round and approximately 200 volts. This will stop when short is removed across R18.

14. Remove test load resistor connected in previous step and note high voltage pulse jumping about 1/8″ and sustaining a healthy arc.

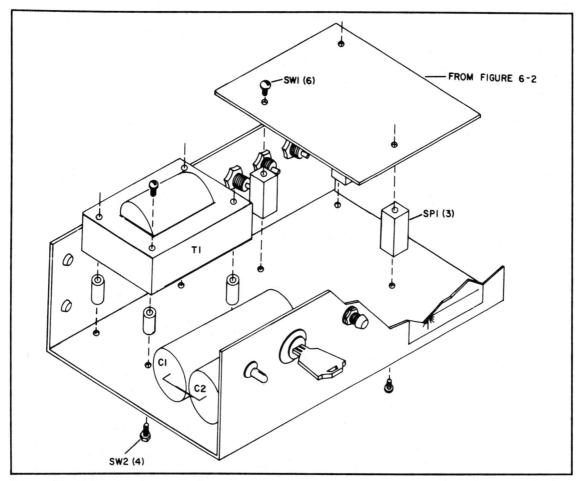

Fig. 6-6. Mounting major components to case.

SPECIAL NOTE ON BALLASTING YOUR HE-NE TUBE

There are many tubes on the market today in good to poor operating condition. Our ballast resistance values are based on new Aerotech certified laser tubes. These tubes are an excellent choice if the builder is considering purchasing new. You will note that Table 6-2 shows the recommended external ballast for these tubes. Even though there is internal ballasting inside of the unit, external ballasts are recommended and should be placed as near to the anode of the laser tube as possible. When using other tubes of questionable parameters it is suggested to use the formulas given. A load line (Fig. 6-9) is shown for the builder desiring to accurately select his ballast resistors for any in between values of those in Table 6-2. If a tube of totally unknown parameters is used it is suggested to start with a high amount of ballast such as 500 kΩ. If the tube lases, measure the current and further adjust for a 4-5 mA value. If the tube is unstable you may decrease the 500 kΩ to 400 kΩ and repeat above. **Caution: Do not allow any tube to draw over 7 mA for much more than the time it takes to make your measurements.**

Table 6-2. Determination of R_{EX} External Ballast.

INTERNAL BALLAST

0 OHMS AT PORT "NO BAL"
260 kΩ AT PORT ".8 mW"
200 kΩ AT PORT "2 mW"
80 kΩ AT PORT "5 mW"

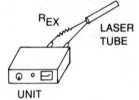

UNIT

Internal ballast values are as noted below and must be adjusted for continuous use using some "external" ballast (R_{EX}) depending on tube used.

Use the following formula for selection of external ballast (R_{EX}) values:

$$(R_{EX}) = \frac{V - E_t}{I_t} - R_{in}$$

V = Voltage selected from Fig. 6-9 at manufacturer's rated tube current
E_t = Voltage across tube from manufacturer's information
I_t = Current through tube from manufacturer's information
R_{in} = Internal ballast at port selected

The following are examples for the Aerotech laser tubes (note line volts to unit = 120 Vac)

Note the wattage of resistors designated for R_{EX}. High values of ballast may be eliminated by varying the input voltage to the unit via a variac.

TUBE	PWR	V	E_t	I_t	R_{in}	R_{EX} External Ballast	R_m*	FORMULA
*LT05R	0.8 mW	3550	1150	4 mA	260 kΩ	340 kΩ/10 W	80 kΩ	above
*LT02R	2 mW	3475	1500	5 mA	200 kΩ	195 kΩ/5 W	80 kΩ	above
*LT04R	5 mW	3425	2100	6 mA	80 kΩ	140 kΩ/7 W	80 kΩ	above

*Do not go below R_m manufactures recommended values for ballast. This value is inclusive of R_{in} + R_{EX}

88

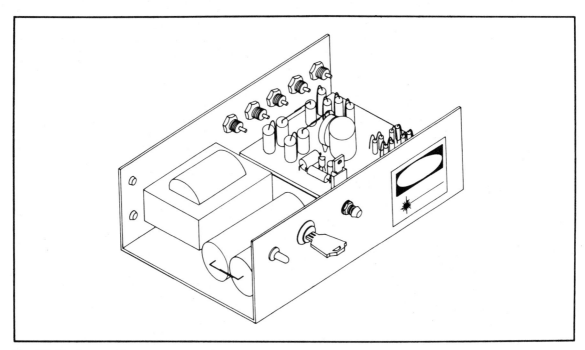

Fig. 6-7. Final assembly.

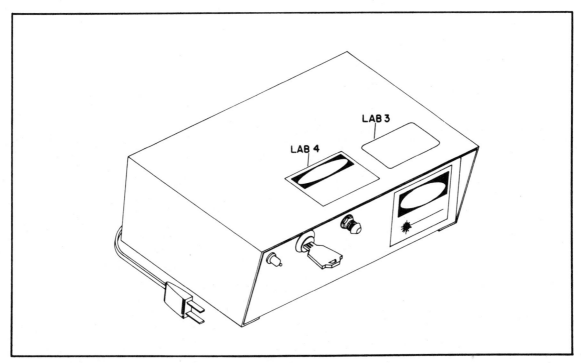

Fig. 6-8. Completed assembly.

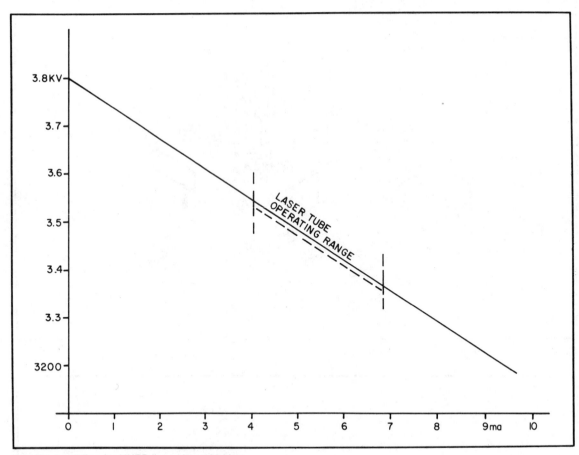

Fig. 6-9. Load line for HNES line volts = 120 Vac.

You are always on safe ground with high ballast and the recommended tube current because line and circuit fluctuations are more absorbed in the high ballast than the tube.

Chapter 7

Voice Modulation of a He-ne Laser (HNM1)

THE FOLLOWING PROJECT SHOWS HOW TO ACTUALLY TRANSMIT YOUR VOICE and other audio information via a helium-neon laser beam. Operation requires other equipment such as our Laser Light Detector (LLD3) described in Chapter 8 for receiving and demodulating these signals.

A helium-neon laser system such as our HNE3 described in Chapter 6 is shown retrofitted with the modulation circuitry as shown in Fig. 7-1. The degree of percentage of modulation will depend on the ability of the laser tube to remain conducting when in the negative half of the modulation envelope. Because the modulation is amplitude variant, a definite current swing varies both positive and negative around the quiescent dc operating value of the tube. Most tubes in this power range usually operate at about 5 mA with limits of plus and minus 0.5 mA. This corresponds to a theoretical modulation percentage of 20% if the actual output intensity is proportional to the current swing. This is not always the case, however, successful communication can be accomplished over considerable distances both in free air and also using fiberoptics.

It is clear that the resultant modulated signal received at the detector is dependent on the ability of the laser tube to vary in intensity along with the changing current thru it. This places certain degrees of merit on the particular laser tube selected and can vary appreciably.

All laser tubes will work to some extent but some work better than others. *Information Unlimited* has selected certain tubes that are more suited for this application. They are available on special request.

The problems encountered with this system are moderate hum and noise on the beam along with the saturating of the receiver-detecting phototransistor at close ranges.

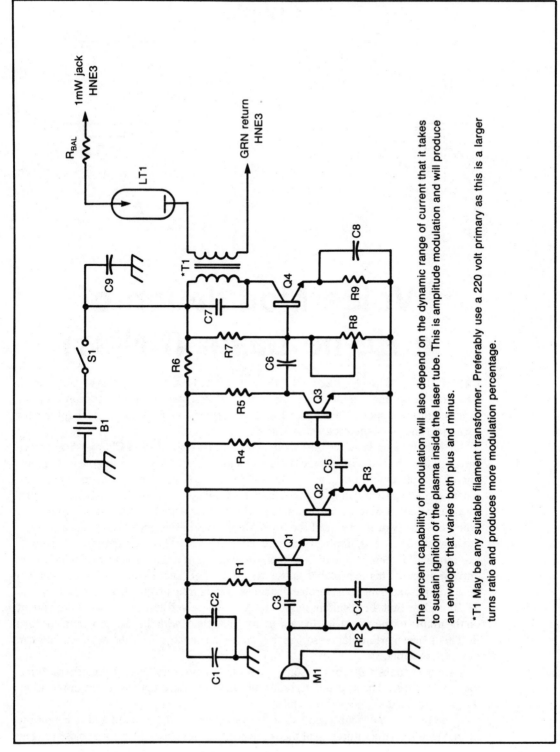

The percent capability of modulation will also depend on the dynamic range of current that it takes to sustain ignition of the plasma inside the laser tube. This is amplitude modulation and will produce an envelope that varies both plus and minus.

*T1 May be any suitable filament transformer. Preferably use a 220 volt primary as this is a larger turns ratio and produces more modulation percentage.

Fig. 7-1. Modulator schematic.

CIRCUIT THEORY

A high impedance crystal mike (M1) as capacitively coupled by (C3) to a Darlington pair consisting of transistor (Q1, Q2). See Fig. 7-1.

This circuit produces an input impedance of approximately the value of R3 times the beta squared of Q1 and Q2 and is necessary for proper terminating of M1. You will note C4 across the bias resistor R2 for prevention of high-frequency oscillation.

Signal voltage across R3 is approximately equal to that at the output of M1 but now is produced across a lower impedance for coupling to the amplifier (Q3). This circuit is an emitter follower and now amplifies the signal voltage necessary for driving the final output transistor (Q4). The collector of Q4 is transformer coupled via T1 that further steps up the signal voltage via its turn ratio. The circuit as shown is battery powered and is completely dc isolated from the laser power supply via T1. A decoupling resistor (R6) and storage capacitor (C1) help to stabilize the circuit.

CONSTRUCTION

For circuit layout follow Fig. 7-1 and Table 7-1. A small perfboard is recommended and assembly should follow audio frequency wiring techniques. Interconnecting leads to the actual laser power supply should be kept as short as possible to avoid hum, noise, etc. Keep all leads direct, avoid wire bridges, note polarity of components and use actual leads wherever possible for connection.

OPERATION AND ADJUSTMENT

1. Obtain a working light detector and associated headsets as described in Chap-

Table 7-1. Parts List for the HNM1.

R1	1	2.2 MΩ 1/4-watt resistor	
R2	1	1 MΩ 1/4-watt resistor	Complete kit of
R3	1	1 kΩ 1/4-watt resistor	parts available from
R4	1	390 kΩ 1/4-watt resistor	Information Unlimited
R5,6	2	470 Ω 1/4-watt resistor	P.O. Box 716
R7	1	4.7 kΩ 1/4-watt resistor	Amherst, NH
R8	1	2 kΩ trimpot resistor	03031
R9	1	27 Ω 1/2-watt resistor	
C1,8	2	100 µF at 25-volts electrolytic cap	
C2,4	2	470 pF 50-V disc cap	
C3	1	.001 µF 50-V disc cap	
C5,6	2	10 µF at 25-V electrolytic cap	
C7	1	.1 µF 50-V disc	
C9	1	1000 µF at 25-V electrolytic cap	
Q1,2,3	3	PN2222 NPN GP transistor	
Q4	1	D40D5 NPN power tab transistor	
T1	1	12-volt to 220-volt small power transformer (see text in Fig. 7-1).	
B1	1	1.5-volt AA Cells for 12 volts	
S1	1	Small on/off switch	
R BAL	1	Added ballast resistance (note text)	
LT1	1	Laser tube (note text)	
PB1	1	3 × 6″ piece of perforated circuit board	
BH1	1	8 AA cell battery holder for 12 volts	

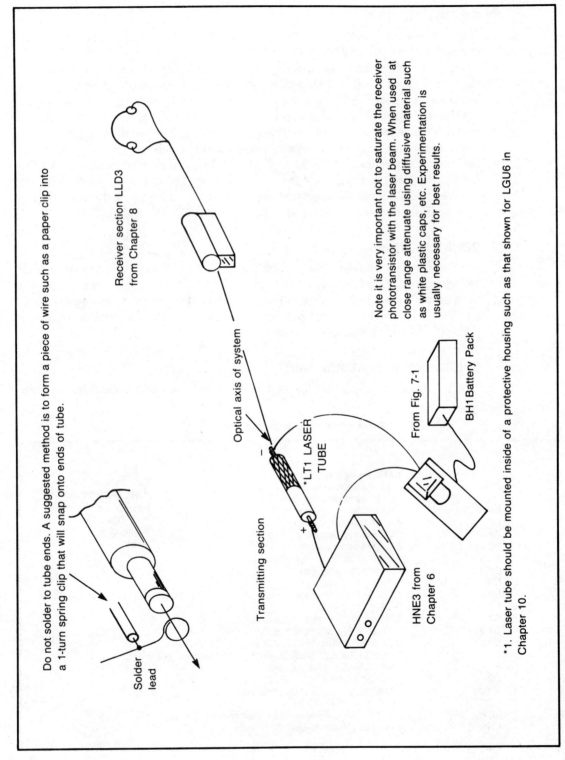

Do not solder to tube ends. A suggested method is to form a piece of wire such as a paper clip into a 1-turn spring clip that will snap onto ends of tube.

Solder lead

Receiver section LLD3 from Chapter 8

Optical axis of system

Transmitting section

*LT1 LASER TUBE

From Fig. 7-1

BH1 Battery Pack

HNE3 from Chapter 6

Note it is very important not to saturate the receiver phototransistor with the laser beam. When used at close range attenuate using diffusive material such as white plastic caps, etc. Experimentation is usually necessary for best results.

*1. Laser tube should be mounted inside of a protective housing such as that shown for LGU6 in Chapter 10.

Fig. 7-2. One-way laser system.

ter 8. Verify operation. Note that the laser beam easily saturates the photodetector transistor of this device and must be heavily attenuated when used at close ranges. This can be accomplished by filters, diffusion plates, etc. The system in the lab when tested, was readable when completely out of the field of the main beam. This means that the range capability is considerable owing to the fact of the minute light required to produce a readable signal. The best approach here is to experiment for the best results.

2. Obtain a working laser power supply such as that described in Chapter 6. Connect a 1 mW He-Ne laser tube and adjust circuit to 5 mA by selection of external ballast resistors. Note that optimum performance will require a special tube intended for modulation and it is available through Information Unlimited, P.O. Box 716, Amherst, NH 03031. Write or call (603) 673-4730 for prices and availability.

3. Verify operation of modulation circuit and connect secondary of T1 in series with the common lead of the power supply and the cathode of the laser tube. Apply power to the modulator circuit via switch S1 and adjust R8 for a current of 25 mA as noted when meter is connected in series with one of the battery contacts.

4. Align and test at various ranges. See Fig. 7-2.

Chapter 8

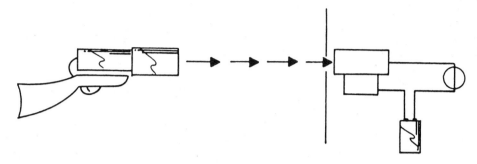

Optical
Light Detector
and Receiver (LLD3)

THIS PROJECT SHOWS HOW TO CONSTRUCT OUR ELECTRO-OPTICAL RECEIVER that is capable of detecting and reproducing the modulated information placed on an optical beam of energy. It allows listening to any varying periodic source of light such as calculator displays, TV sets, normal lighting, the light produced from a fire, lightning, IR sources, etc., and of course intentionally modulated beams for voice or other analog communications.

High speed, short duration optical pulses such as those from solid-state injection lasers may be detected with limited success. It is suggested to use the detector described in Chapter 11 for this application.

A very capable source of modulated laser light is that described in Chapter 7, where the continuous beam of a low-powered helium-neon laser is amplitude modulated at around audio frequency to about a 15% to 25% level. When used together, communications can range over several kilometers. Other uses of this project are the receiver section for a long-range intrusion detector, perimeter protection, announcer, light level indicator, etc.

CIRCUIT DESCRIPTION

Your optical light detector (Fig. 8-1 and Table 8-1) utilizes a sensitive phototransistor (Q5) placed at the focal point of a lens (LE2). The output of Q5 is fed to a sensitive amplifier consisting of an array (A1) and is biased via the voltage divider consisting of R14 and R1. The base is not used. Q5 is capacitively coupled to a Darlington-pair for impedance transforming and is further fed to a capacitively-coupled cascaded pair of common-emitter amplifiers for further signal amplification.

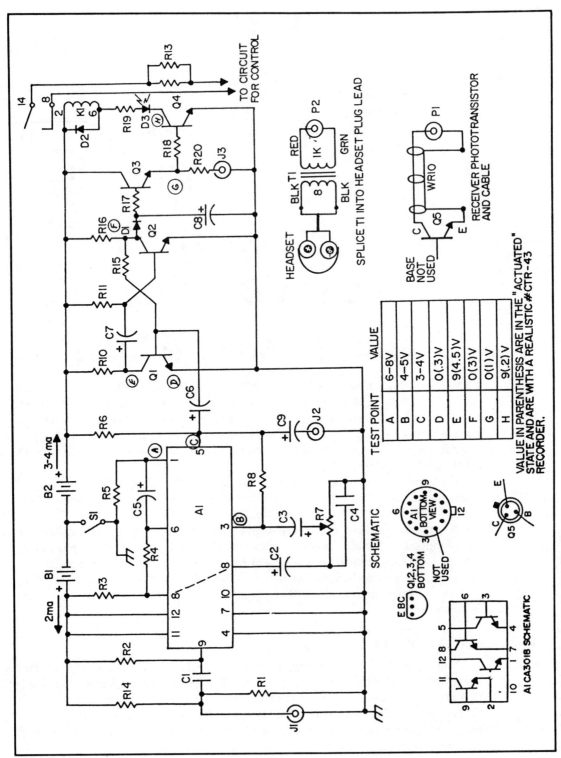

Fig. 8-1. Optical light detector schematic diagram.

Table 8-1. Parts List for the Optical Light Detector (LLD1).

R1,4,8	3	390 kΩ 1/4-watt resistor
R2	1	5.6 MΩ 1/4-watt resistor
R3,5,6	3	6.8 kΩ 1/4-watt resistor
R10,16	2	5.6 kΩ 1/4-watt resistor
R11,14,15 17	4	100 kΩ 1/4-watt resistor
R13 A&B	2	10 Ω 1/4-watt resistor
R18	1	220 Ω 1/4-watt resistor
R19	1	100 Ω 1/4-watt resistor
R7S1	1	5 kΩ pot & switch combination
C1	1	.047 ΩF, 25 V disc cap
C2,3,5 6,7,9	6	1 to 3 μF, 25 V elect
Q1,2,3,4	4	PN2222 NPN silicon transistor
Q5	1	Phototransistor L14G3 or equivalent
A1	1	CA3018 amp array
C1	1	1N914 signal diode
D2	1	N4007 1000 V diode
D3	1	LED indicator FLV106
K1	1	Mini dip relay 6 V spst
J1,2,3	3	RCA phono jacks
P1,2	2	RCA phono plug
CL1,2	2	Battery snap clips for 9 V rect.
CA1	1	Case 4 × 2-1/8 × 1-5/8 aluminum mini box
*PB1	1	Perfboard 1-1/4 × 2-5/8 Printed circuit board available
KN1	1	Small plastic knob for R7/S1
BU1	1	3/8″ plastic bushing for D3
BU2	1	Cord clamp bushing for wire from K1
LE2	1	Lens 54 × 89mm
WR1	12″	Shielded mike cable
WR4	24″	24″ plastic hook-up wire
EN2	1	Enclosure 6-1/2 × 2-3/8 sked 40 PVC (fabbed as shown)
CA3,4	2	2-3/8″ plastic caps
DO1	1	Mtg dowel 2 × 2 soft wood
FTR1	1	2″ IR filter
B1,2		9 V transistor battery

Optional Parts Available Individually

*HS1		8-ohm microphone headphone with T1 matching transformer & P2 plug (Assembled) (Per inset Fig. 8-1)
*PCHGA1		Printed circuit board—rework (Per Fig. 8-3)

All parts and assemblies available from Information Unlimited, P.O. Box 716, Amherst, NH. Write or call (603) 673-7421 for price and availability.

Parts marked with an asterisk are individually available.

Sensitivity control R7 controls base drive to the final transistor of the array and hence controls overall system sensitivity. Output of the amplifier array is capacitively coupled to a one shot consisting of Q1 and Q2 in turn integrating the output pulses of Q2 onto capacitor C8 through D1. This dc level now drives relay drivers Q3 and Q4 activating K1 along with energizing indicator D3, consequently controlling the desired external circuitry. The contacts of K1 are in series with low-ohm resistors R13 to prevent failure when switching capacitive loads.

J2 allows "listening" to the intercepted light beam via headsets. This is especially useful when working with pulsed light sources such as GaAs lasers or any other varying periodic light source.

CONSTRUCTION STEPS

1. Identify all components. Note that indicated layout must be followed for proper performance. See Figs. 8-2 and 8-3.

2. Fabricate an aluminum mini box as shown for J1, J2, BU1, BU2, and R7/S1. Assemble these parts in place as shown. Note ground lug under J1. Note mating holes for securing Q5 housing (Fig. 8-4).

3. Reference corner of assembly board and mark as shown in Fig. 8-2.

4. Insert CA3018 and carefully position as shown. Bend over several leads to keep it from falling out. This insertion of the CA3018 may require several attempts before aligning the 12 leads with the appropriate holes in the perfboard or PC board as shown. note C5 and R5 shown mounted under the CA3018. Use Figs. 8-2 and 8-3 when using a printed circuit board.

5. Using standard audio frequency wiring techniques, proceed to wire and solder starting with C1, inserting the designated components and soldering point-by-point. Carefully check for accuracy and the quality of all solder joints. Remember mistakes can ruin the CA3018. *Observe polarity of electrolytic capacitors and position of relay.*

6. Attach battery clips by inserting leads through holes in perfboard adjacent to their termination points as shown. This method strain relieves these wires.

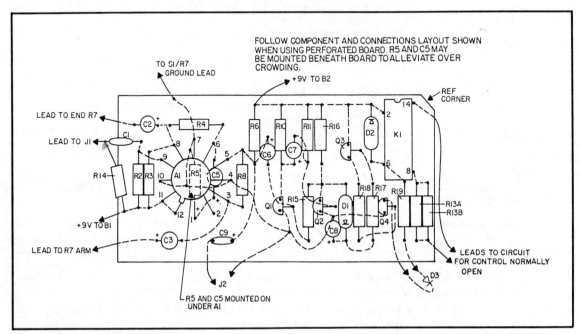

Fig. 8-2. Printed circuit board layout.

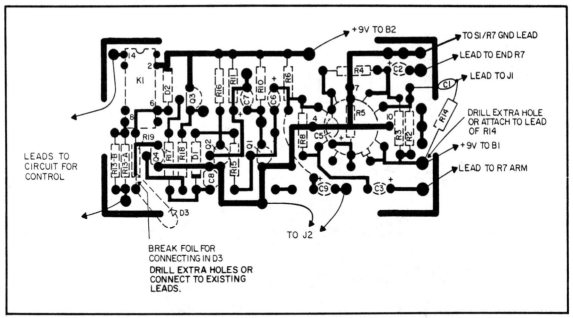

Fig. 8-3. Printed circuit board foil side.

7. Connect C4 across R7 as shown. Connect ground end of R7 to ground lug J1 via uncut lead from C4. Connect buss jump from this point to one of the lugs of S1 as shown. Connect R1 between J1 and the ground lug of J1 with above and solder this point. Note these leads must be as short and as direct as possible to prevent noise and hum.

8. Note that the assembled board should have the following leads for connection to the components in the aluminum mini box.

☐ Input lead of C1 along with R14 for connection to J1 (short as possible) use component lead.

☐ Ground lead from pins 4, 7, and 10 of CA3018 for connecting to ground of case at S1/R7.

☐ Two buss leads to R7 end and R7 arm (use uncut leads of components if possible). These leads must be as short as possible.

☐ Leads to D3. Note foil must be broken on PC board as shown for inserting D3 in series.

☐ Leads to J2. Use C9 lead if possible.

☐ Control leads from K1.

9. Make a visual check for faulty soldering, wiring errors, and shorts.

10. Fabricate EN2 from a 6-1/2" piece of 2-3/8" OD schedule 40 PVC tubing. Drill a 3/4" hole as shown approximately 3" from rear end. This hole is for optical alignment. Note two mating holes for securing to CA1. These holes are also in top of this piece of access with a screwdriver. The bore axis of this tube should be parallel with CA1. It may be more convenient to slot the rear hole to allow slight side movement

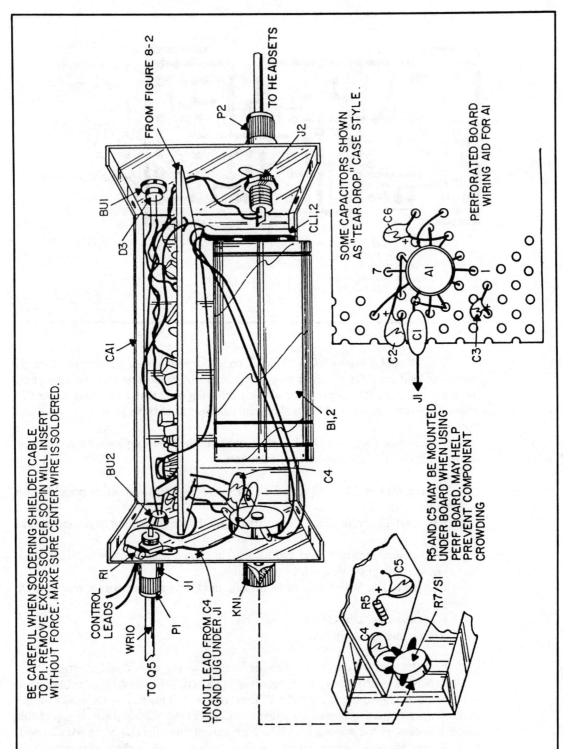

Fig. 8-4. Assembly.

for final alignment. The large 3/4″ hole can be covered with a plug, tape, etc., when not needed.

11. Fabricate DO1 centering dowel from a 2″ length of 1-1/2″ OD (or thereabouts) for a smooth sliding fit into EN2. Q5 is mounted for optical centering, via small pin holes in the wood for securing via its leads. Cable WR1 is fed to Q5 via a slight off-center feed hole in the dowel. Connection is made by soldering to exposed leads of Q5 (watch for overheating) and then securing with RTV or equivalent. Leads to Q5 should be left long enough to allow touch up repositioning to true optical axis for final alignment.

12. Fabricate CA3 and CA4 from a 2-3/8″ plastic cap. Remove end with exception of 3/8″ lip to retain lens (LE2) and optional filter (FTR1) against end of EN2. CA4 is also a 2-3/4″ plastic cap. Place small hole for cable (WR1). Hole should create friction hole to prevent DO1 from sliding once set. See Fig. 8-5.

Secure with RTV when complete and finally aligned. It is assumed that the assembled unit to this point has been wired correctly, with no shorts, and good solder connections.

13. Install P1 into J1.

TESTING

1. Turn R7/S1 full ccw (off position).

2. Connect one terminal of a fresh 9-volt battery to CL1 and connect a milliammeter between the unused contact of the battery and the clip. Turn on R7/S1 and note current reading of approximately 2 mA. Fully connect battery and designate B1.

3. Repeat above using second battery designated B2. Note current reading of 3-4 mA. Turn up gain R7 and note B2 current increasing to 12 mA and D3 lighting when light is detected. This is the relay current and indicates on "on" state. Current should drop back to 3 mA when relay turns "off" in several seconds. Note that R7 must not be set at too high a gain or the unit will not turn off. R7 will have to be set way down in normal background light if the unit is used without FTR1.

4. Adjust Q5 to focal length of LE2. This is accomplished by pointing the unit at a distant source of light and placing a piece of paper over Q5. Adjust DO1 position for a sharp image of light source over Q5 lens. This is easily accomplished through the access hole in the enclosure.

5. Plug a high-impedance headphone into J2. Note Fig. 8-1 showing standard 8-ohm headsets with spliced-in matching transformer for stepping up to 1000 ohms. High impedance headsets are scarce and usually uncomfortable to wear.

6. Turn on R7/S1 and slowly turn up gain until a loud 60-cycle hum is heard. This is the normal lighting frequency being picked up by Q5 and at normal ambient light conditions will completely block the amplifier. Reduce the gain and attempt to point Q5 at various objects indicating different levels of signal depending on the reflection characteristics of the surfaces, etc. Point at TV screen, scope or any periodic changing source of light. You will note that the circuit is relatively prone to power line hum pick up. This is because of the plastic enclosure EN2 and the floating base lead of Q5. This may be biased with a resistor to the emitter for use with high signal levels. It is assumed that testing will be done in normal electrical lighting for this step. If not, you

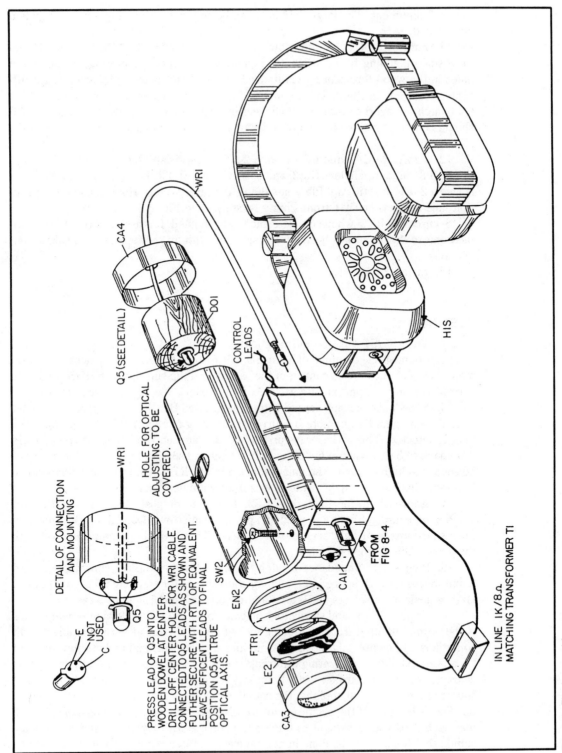

DETAIL OF CONNECTION AND MOUNTING

E NOT USED C

Q5

WRI

PRESS LEAD OF Q5 INTO
WOODEN DOWEL AT CENTER.
DRILL OFF CENTER HOLE FOR WRI CABLE
CONNECTED TO Q5 LEADS AS SHOWN AND
FUTHER SECURE WITH RTV OR EQUIVALENT.
LEAVE SUFFICENT LEADS TO FINAL
POSITION Q5 AT TRUE
OPTICAL AXIS.

HOLE FOR OPTICAL
ADJUSTING. TO BE
COVERED.

WRI

CA4

Q5 (SEE DETAIL)

DO1

CONTROL
LEADS

HIS

SW2

EN2

FTR1

LE2

CA3

CA1

FROM
FIG 8-4

IN LINE 1K/8Ω
MATCHING TRANSFORMER T1

Fig. 8-5. Final assembly.

THE LLDI LIGHT DETECTOR IS INSTALLED AS THE TARGET WITH ITS LENS BEING APERTURED BY SUCCESSIVELY SMALLER COVERS. THESE COVERS ARE REPLACED AS THE USER BECOMES MORE ACCURATE. RANGE MAY BE UP TO 20 METERS WITH THE LHP2 SIMULATED LASER, OR UP TO SEVERAL HUNDRED METERS USING THE GENUINE Ga As LASER SUCH AS OUR LRG3, LP3, OR SSLI. USE INGENUITY IN SIMULATING THE DEVICE INTO A GUN CONFIGURATION. TRIGGER IS OBVIOUSLY THE SWITCH.

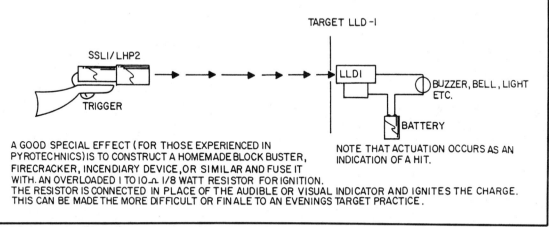

A GOOD SPECIAL EFFECT (FOR THOSE EXPERIENCED IN PYROTECHNICS) IS TO CONSTRUCT A HOMEMADE BLOCK BUSTER, FIRECRACKER, INCENDIARY DEVICE, OR SIMILAR AND FUSE IT WITH. AN OVERLOADED 1 TO 10 Ω 1/8 WATT RESISTOR FOR IGNITION. THE RESISTOR IS CONNECTED IN PLACE OF THE AUDIBLE OR VISUAL INDICATOR AND IGNITES THE CHARGE. THIS CAN BE MADE THE MORE DIFFICULT OR FINALE TO AN EVENINGS TARGET PRACTICE.

NOTE THAT ACTUATION OCCURS AS AN INDICATION OF A HIT.

Fig. 8-6. Laser shooting gallery.

THE LLDI IS USED AS THE RECEIVER FOR A TIGHT BEAM OF LIGHT FROM THE LHP2 SIMULATED LASER OR THE GaAs DEVICE REFERENCED IN FIG 6-6 USED AS A LIGHT TRANSMITTER WHEN THE BEAM IS INTERRUPTED BY A FOREIGN BODY, AN ALARM IS SET OF AND CAN BE CONNECTED TO REMAIN IN AN ON STATE UNTIL THE SYSTEM IS RESET. THE ALARM CAN ALSO BE MADE TO SOUND ONLY WHEN THE BEAM IS ACTUALLY INTERRUPTED.

YOUR LLDI LIGHT DETECTOR WITH YOUR HEADPHONES ALLOWS YOU TO LISTEN TO ANY CHANGING LIGHT SOURCE, SUCH AS TV SCREENS, LIGHTS, FIRES, LIGHT BEAM COMMUNICATOR, TRANSMITTERS, ETC.

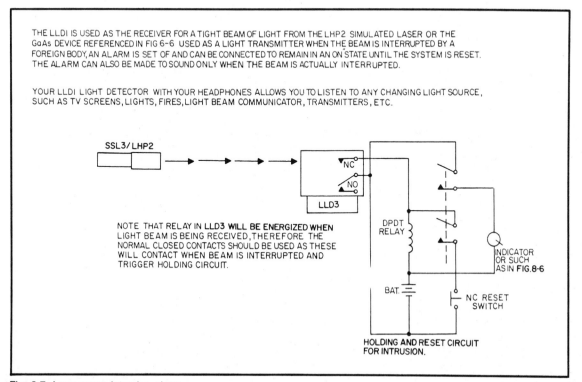

NOTE THAT RELAY IN LLD3 WILL BE ENERGIZED WHEN LIGHT BEAM IS BEING RECEIVED, THEREFORE THE NORMAL CLOSED CONTACTS SHOULD BE USED AS THESE WILL CONTACT WHEN BEAM IS INTERRUPTED AND TRIGGER HOLDING CIRCUIT.

HOLDING AND RESET CIRCUIT FOR INTRUSION.

Fig. 8-7. Long range intrusion alarm.

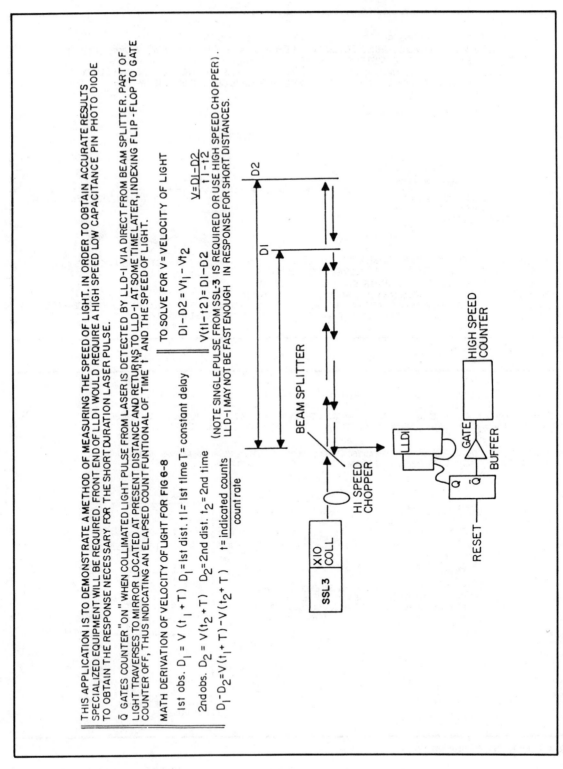

Fig. 8-8. Light speed indicator.

may not obtain the 60-Hz hum from the varying light. If you troubleshoot a faulty circuit, it may be convenient to use the test points shown in Fig. 8-1 and thoroughly familiarize yourself with the *Circuit Description* section given at the beginning of this project. See Figs. 8-6 through 8-8.

Chapter 9

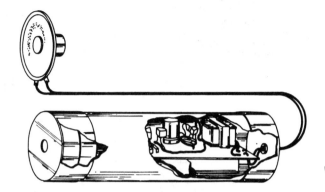

Laser Light Detector/Target (LDT1)

THE LTD1 LASER LIGHT DETECTOR TARGET PROJECT IS INTENDED TO SOUND AN alarm whenever the light level through the aperture exceeds a certain value. (See Figs. 9-1, 9-2 and Table 9-1.) The threshold level is set by adjustment of R2. The device is designed to be used in conjunction with any laser light gun or other similar device. It can provide a means of improving hand gun steadiness by holding the laser beam on the aperture as long as possible. A means for instinctive point firing of such weapons as shotguns and machine guns is also realized.

The circuit may be modified so that the beam is required to keep the alarm "off" thus providing an intrusion link, or as a motion detector where an object when misaligned would cause an alarm state. Obviously, there are many uses for this unit and many are left to the ingenuity and discretion of the user.

DEVICE DESCRIPTION

The device is shown housed in a plastic cylindrical tube (EN1) fitted with plastic cap (CA1), an aperture cap (AP1) and diffusion plate (DP1). You will note that the aperture cap AP1 is intended to have its opening vary in size depending on range and other parameters of the light source. The DP1 diffusion plate is made from two plastic caps assembled together as shown Fig. 9-3. This method is necessary when being used with the highly coherent beam of light from a laser because total diffusion is necessary.

CIRCUIT DESCRIPTION

Light entering the aperture cap AP1 is diffused by DP1 diffusion plate. Phototran-

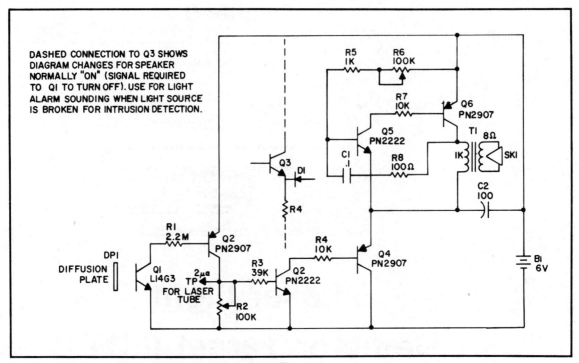

Fig. 9-1. LDT1 electronic laser target wire for normal "off" mode.

sistor (Q1) is placed so that it picks up a small amount of the diffused light and causes its collector to conduct through the base resistor (R1) for the transistor (Q2). Q2 collector causes a voltage to be developed across the sensitivity adjust pot (R2) proportional to its setting. This voltage now turns on the transistor (Q3) through the base resistor (R3). Q3 turns on clamp transistor (Q4) through the resistor (R4). Q4 now acts as a

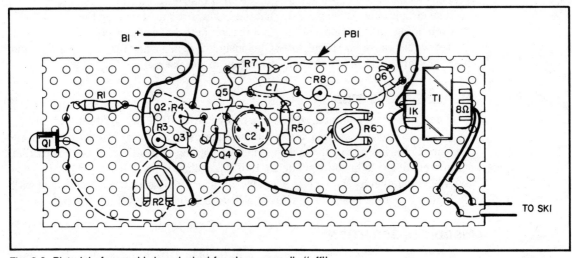

Fig. 9-2. Pictorial of assembly board wired for alarm normally "off".

110

Table 9-1. Laser Target Parts List (LDT1).

R1	1	2.2 MΩ 1/4-watt resistor
R2,6	2	100 kΩ trimpot resistor
R3	1	39 kΩ 1/4-watt resistor
R4,7	2	10 kΩ 1/4-watt resistor
R5	1	1 kΩ 1/4-watt resistor
R8	1	100 Ω 1/4-watt resistor
C1	1	.1 μF 50 V disc capacitor
C2	1	100 μF/25 V elect. capacitor
*Q1	1	L14G3 phototransistor
Q2,4,6	3	PN2907 PNP transistor
Q3,5	2	PN2222 NPN transistor
T1	1	1000-Ω to 8-ohm transformer
SK1	1	8-ohm 2″ speaker
D1	1	1N914 signal diode
DP1	1	Diffusion plate (use 2 plastic caps, see text 1-1/2″)
AP1	1	Aperture cap 1-7/8″
CA1	1	Rear cap 1-7/8″
BU1	1	Small wire bushing
EN1	1	Enclosure fab from 7″ of 1.9″ diameter plastic tube
PB1	1	Perfboard 1-1/2″ × 3-3/4″
WR1	1	Hook up wire use #24
B1	1	9-V battery

LDT1 Laser Target Kit is available from Information Unlimited, P.O. Box 716, Amherst, NH 03031, write or call (603) 673-4730 for price and availability. Parts marked with an asterisk are individually available.

switch grounding all the return points of the transistor audio oscillator (Q5 and Q6). The tone of the oscillator is controlled via trimpot (R6). Audio signal output is taken via output transformer (T1) and fed to speaker (SK1). Feedback to the system is via resistor (R8) and capacitor (C1). The above circuit is silent or nonactivated without a light signal. This may be reversed by changing the circuit as shown in the Fig. 9-1 inset. The output voltage across R2 when turning on Q3 now applies a positive voltage to Q4 causing it to turn off or ungrounding the oscillator. The diode (D1) provides a positive turn-on of Q4. This circuit now will sound the alarm until light is detected, which is opposite to the previous design.

CONSTRUCTION STEPS

1. Layout and identify all parts. Select mode of operation from Fig. 9-1.

2. Assemble board as shown in Fig. 9-2. Note PNP and NPN transistors. Use leads of components as connecting wires whenever possible. Avoid bare wire bridges and short-circuit points.

3. Fabricate DP1 from two milky white plastic caps placed next to one another as shown in Fig. 9-3.

4. Fabricate EN1 from a 7″ length of 1.9″ tubing.

5. Fabricate AP1 aperture and CA1 rear cap., as shown in Fig. 9-3.

6. Final assembly is as shown in Fig. 9-4.

7. Check using a high intensity light source such as a helium-neon laser. Lower

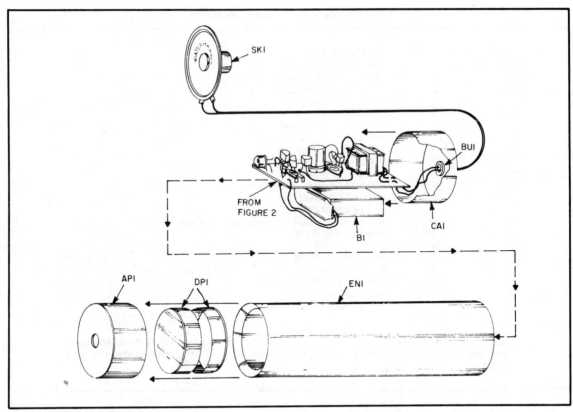

Fig. 9-3. Blow up.

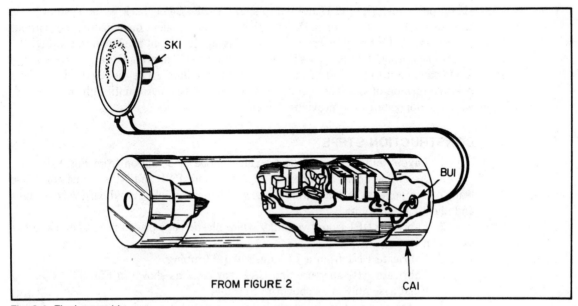

Fig. 9-4. Final assembly.

intensity sources may be used by opening up the aperature hole in AP1. The unit is desensitized by the adjustment of R2. The tone of the oscillator is adjusted by R6.

8. Mount the assembly as desired for your particular application. If used for an intrusion detection the unit must be sturdily mounted.

Chapter 10

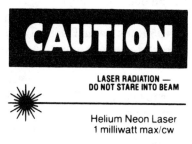

CAUTION

LASER RADIATION —
DO NOT STARE INTO BEAM

Helium Neon Laser
1 milliwatt max/cw

CLASS II LASER PRODUCT

Visible Red 0.3-2 mW Continuous Laser Gun (LGU6)

T HE FOLLOWING PROJECT SHOWS HOW TO CONSTRUCT A PORTABLE SELF-CON-
tained device capable of projecting a visible red laser light beam at 6328 Å. The
property of this light is very high spatial and temporal coherence with a color temperature equivalent of many millions of degrees. This property allows projection of this
energy for great distances with little beam divergence. As an example, one could easily note a brilliant spot when the device is projected on a reflective surface such as
a road sign for distances of one mile or more. The unit described here is built in a ri-
fle/pistol configuration. It contains its own rechargeable batteries and power supply
and weighs less than two pounds. When used without the collimator, the unit is only
about 12″ long, hence a pistol configuration, when used with the 2 mW tube it is about
19″ long, hence a rifle configuration.

The unit is shown constructed around a 1 and 2 mW laser tube whose output is
at 6328 Å with a beam divergence of 1.3 milliradians. The beam width is about .65
mm. When collimated the divergence decreases and the beam width increases by the
magnification factor of the optical system used.

A shorter version of this unit can be built housing the batteries in an enclosure
and attaching them to your belt, etc. The unit also works well on a vehicle's 12-volt
system and can be adapted to a cigarette lighter plug etc.

THEORY OF OPERATION

This device, in principle, is not complicated in theory or operation. (It would be
very critical in adjustment, construction and operation if the tube was to be made from
scratch).

The basic device is very simple, that of a gas-discharge tube that is highly evacuated and then filled with gas, placed between two mirrors forming a resonant optical cavity. When the gas is excited via an external energy source such as a current discharge, photons are produced and due to the amplifying action of the cavity and mirrors, laser radiation is produced. A more comprehensive explanation of laser theory is explained in Chapter 11 (RUB5 Ruby Laser Plans). This sounds simple, however, certain limiting parameters and obstacles are present and must be reduced or eliminated to obtain any decent performance.

A helium-neon laser is a poor amplifier as lasers go, consequently all the following efforts must be made if lasing is to take place. The gas discharge tube must be properly made and sealed for a high vacuum, impurities cannot exist, gas mixture must be pure and applied at the correct pressure. Mirrors must be of the dielectric type for maximum optical efficiency (approaching 100%). Optical alignment must be nearly perfect. Once these demands are fulfilled the following takes place; atoms of gas are excited by an electric discharge to an energy state where photons of coherent light are produced upon these atoms returning to a stabler state. Much of this external energy is wasted in exciting atoms whose photons do not contribute to the laser beam. Some of these photons, however, manage to be reflected back and forth by the mirror and consequently stimulate other excited atoms to do the same, thus producing the necessary synchronized wavelets increasing the beams intensity several percent on each pass between the mirrors.

Placement of these mirrors must be in the same plane for nearly perfect transmissions. They must be of the dielectric type which are formed from multiple layers of nonconducting film. This film is formed from transparent material such as the chlorides and fluorides. The thickness of these layers determines attenuation or enhancement of certain frequencies by creating destructive or constructive reinforcing waves. It is this effect that when repeated in many such layers develops nearly 100% reflection at the desired frequency. Note: for those who may consider making their own tubes, care must be taken to mention that regular silver mirrors seldom exceed 95% and tarnish quickly in air. Aluminum coated mirrors are only around 99% reflective. Those losses are sufficient to prevent the device from performing.

CIRCUIT THEORY

The circuitry (See Fig. 10-1 and Table 10-1) consists of a high-voltage transistor inverter. Transistors Q1 and Q2 switch the primary windings of the transformer (T1) via a square wave at a frequency determined by its magnetic properties. Diodes (D1) and (D2) provide base return paths for the feedback current of Q1 and Q2. Resistor R1 limits this base current while R2 provides the necessary electrical imbalance to commence oscillation. The output winding of T1 is connected to a multiple-section voltage multiplier. The multiplier consists of capacitor C2-C3 and diodes D3-D4. This circuit doubles the ac output voltage of T1 to approximately 2000 volts dc across C2-C3. It is this voltage ballasted down to then necessary value for sustaining conduction that actually powers the laser tube. Ballast resistors (R4, R5) limit the current through the laser tube (LT1) to that value specified by the manufacturer, usually around 4 to 5 mA.

Ignition of the laser tube requires a much higher voltage than that necessary for sustaining operation. It is therefore necessary to initially apply 6 to 10,000 volts dc

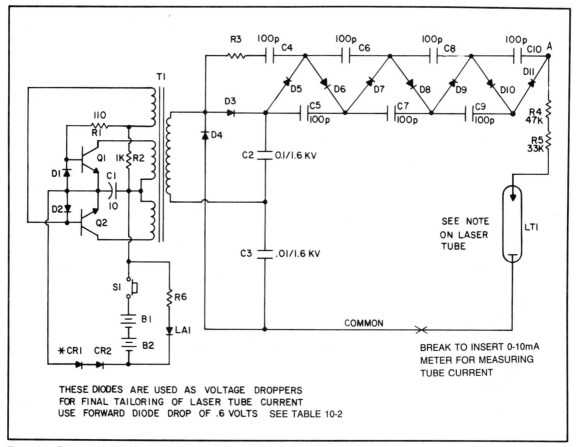

Fig. 10-1. Schematic identification.

to accomplish this. Fortunately ignition of most gas laser tubes of this type requires very little energy. It is only necessary to use a small amount of the available power for this function. Resistor R3 diverts a small amount of energy to the "parasitic" multiplier section consisting of diodes (D5 through D11) and integrating capacitors (C4 through C9). An open-circuit voltage of approximately six times the peak-to-peak output of T1 is produced at the anode of LT1. Upon conduction the parasitic diodes become forward biased by the sustaining current of LT1 and only add a relatively small amount of resistance to the circuit. Note that this parasitic voltage actually adds with the operating voltage developed across C2 and C3 when the system is open circuited (tube not ignited). Resistor R3 should be selected to reliably ignite the tube and yet not be too low a value to consume valuable power as its function ends when ignition is sustained.

The system has been carefully designed to the parameters of laser tubes such as the "LT" series available through Information Unlimited. These are first-rate tubes guaranteed for 20,000 hours (provided they are used within the specified parameters). For the sake of simplification we have selected parameters for most tubes from 0.3 mW to 2 mW, with the only differences being in the voltage of the batteries used. The

Table 10-1. Parts List.

R1	1	110 Ω 2-watt resistor
R2	1	1 kΩ 1/4-watt resistor
R3	1	1 MΩ 1-watt resistor
R4	1	47 kΩ 1-watt resistor
R5	1	33 kΩ 1-watt resistor
R6	1	470 Ω 1/4-watt resistor
C1	1	10 μF at 25 V electrolytic
C2,3	2	.01 μF at 1.6 kV disc
*C4-C10	7	100 pF at 3 kV disc
D1,2	2	1N4002 50 to 100 V power diodes
*D3-D11	9	High voltage diodes 3 kV
*T1	1	Special ferrite transformer (Information Unlimited) Type VI
S1	1	Push-button switch
LA1	1	Yellow or green LED
BU1	1	Small plastic bushing for retaining LA1
**CL1,2	2	Heavy lead battery clip
WR1	24″	#20 600-V vinyl wire or use HV test lead
*PB1	1	3-1/2″ × 1-1/4″ perfboard (optional PC board)
*EN1	1	12″ × 1-5/8″ OD plastic tube (fab Fig. 10-3)
*EN2		or 16-1/4″ of sked 40 2″ PVC for larger LGU6 model (Fig. 10-4)
HA1	1	5-1/2″ × 1-5/8″ OD plastic tube fab Fig. 10-3
BK1	1	6″ × 1/2″ × .035″ soft aluminum fab Fig. 10-7
CA1,2	2	1-1/2″ plastic caps fab Fig. 10-3
CA3,4,5	3	1-5/8″ plastic caps
**BH1	1	8-cell battery holder
**BH2	4	AA cell holder supplied with LGU6
*LAB1	1	Class II warning label for LGU2,4 *** Class III A for LGU6 ***
*LAB2	1	Aperture label
LAB3	1	Certification label
SW1,2,3	3	#6 × 1/4″ sheet metal screws
**B1-8	8	1.2 V Nicad AA batteries
*Charger		Optional charger 50 mA for multiple AA Cells
*COL1K	1	Collimator (optional)
*LT1		See Table 10-3 for selection
CH1	6″	Small beaded chain
Q1,2	2	D4005 NPN Pwr Tab

Complete kit of above or assembled and tested unit is available from Information Unlimited, P.O. Box 716, Amherst, NH 03031. Write or call (603) 673-4730. Parts marked with an asterisk are individually available.

***See below for classification of laser tubes for LT1

LT01R	.3 mW	Class II labeling and compliance
LT05R	.8 mW	Class II labeling and compliance
LT02R	2 mW	Class IIIA labeling and compliance

Note: Buyer beware when purchasing surplus laser tubes. Remember, if they are available at reduced cost, chances are they are defective in one way or another.

**Note when building LGU2 or LGU4 it is only necessary to use the single AA 8-cell holder for 10-12 Vdc. The LGU6 will require 14-16 volts and will use an extra 4-cell holder for a total of 11 to 12 batteries.

system when used with a 0.3 to 0.8 mW tube, uses (8) 1.2 volt AA sized nicads for a total input voltage of 9.8 V. When using a 2-mW tube, it is suggested to use (10) 1.2 volt AA sized nicads for a total of 12 volts. (Certain 2 mW tubes may require (11) 1.2-volt batteries for reliable operation). See Table 10-2 for electrical load line parameters

Table 10-2. Power Supply Load Line.

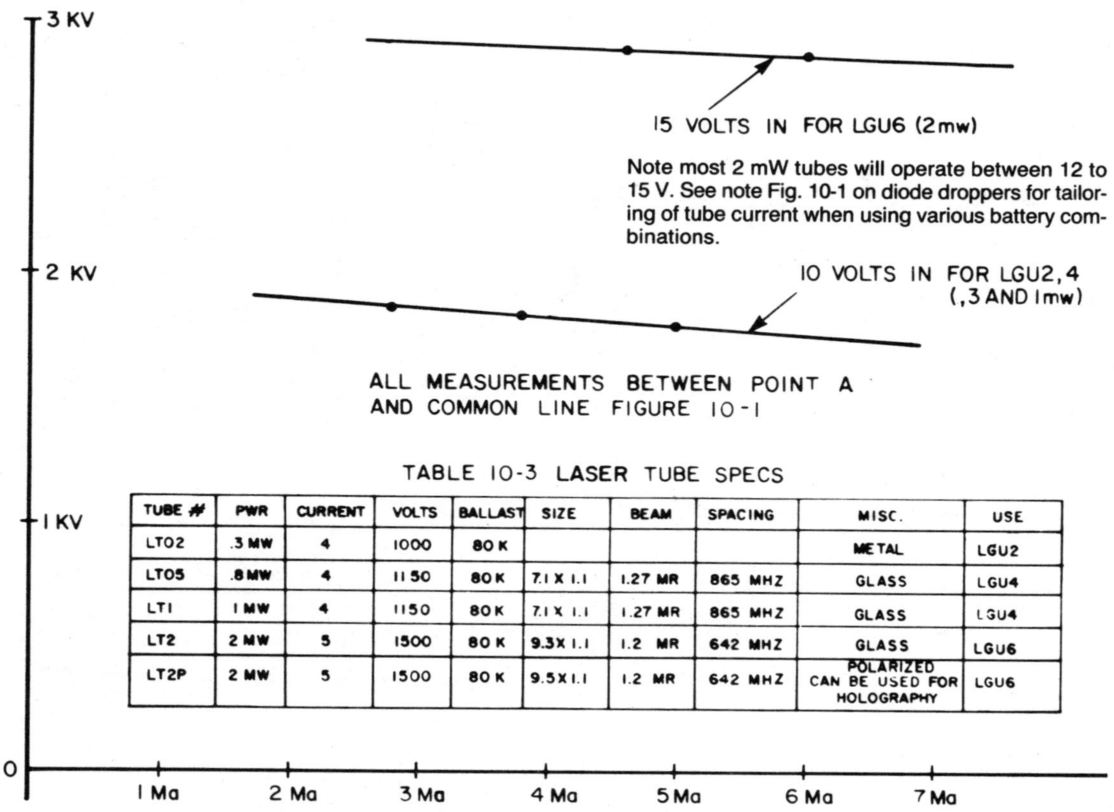

15 VOLTS IN FOR LGU6 (2mw)

Note most 2 mW tubes will operate between 12 to 15 V. See note Fig. 10-1 on diode droppers for tailoring of tube current when using various battery combinations.

10 VOLTS IN FOR LGU2,4 (.3 AND 1 mw)

ALL MEASUREMENTS BETWEEN POINT A AND COMMON LINE FIGURE 10-1

TABLE 10-3 LASER TUBE SPECS

TUBE #	PWR	CURRENT	VOLTS	BALLAST	SIZE	BEAM	SPACING	MISC.	USE
LT02	.3 MW	4	1000	80 K				METAL	LGU2
LT05	.8 MW	4	1150	80 K	7.1 X 1.1	1.27 MR	865 MHZ	GLASS	LGU4
LT1	1 MW	4	1150	80 K	7.1 X 1.1	1.27 MR	865 MHZ	GLASS	LGU4
LT2	2 MW	5	1500	80 K	9.3X1.1	1.2 MR	642 MHZ	GLASS	LGU6
LT2P	2 MW	5	1500	80 K	9.5X1.1	1.2 MR	642 MHZ	POLARIZED CAN BE USED FOR HOLOGRAPHY	LGU6

3 KV

2 KV

1 KV

0

1 Ma   2 Ma   3 Ma   4 Ma   5 Ma   6 Ma   7 Ma

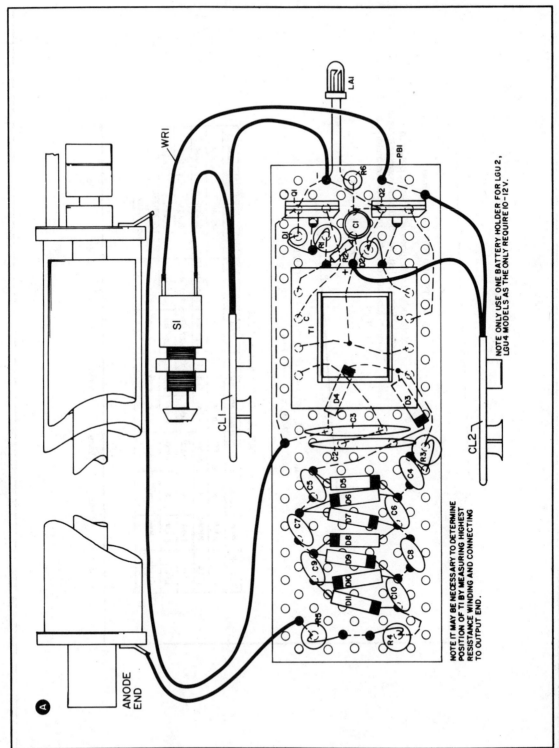

NOTE IT MAY BE NECESSARY TO DETERMINE POSITION OF T1 BY MEASURING HIGHEST RESISTANCE WINDING AND CONNECTING TO OUTPUT END.

NOTE ONLY USE ONE BATTERY HOLDER FOR LGU 2, LGU4 MODELS AS THE ONLY REQUIRE 10-12V.

ANODE END

Fig. 10-2A. Assembly board top view.

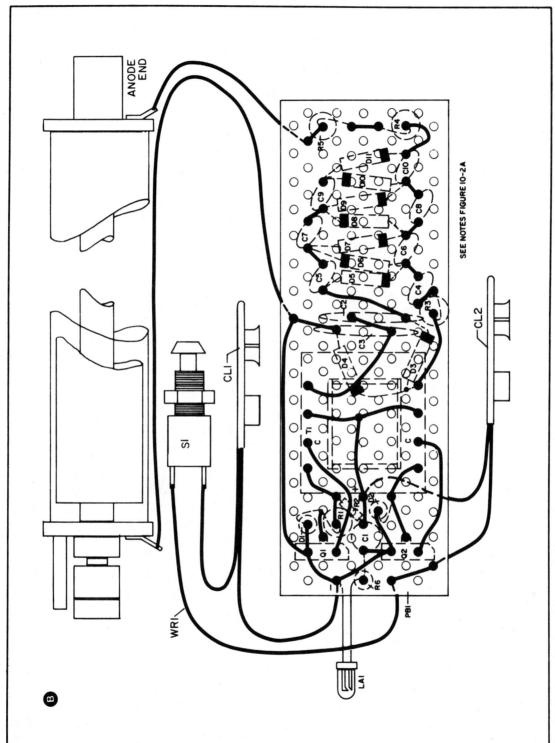

Fig. 10-2B. Assembly board bottom view.

121

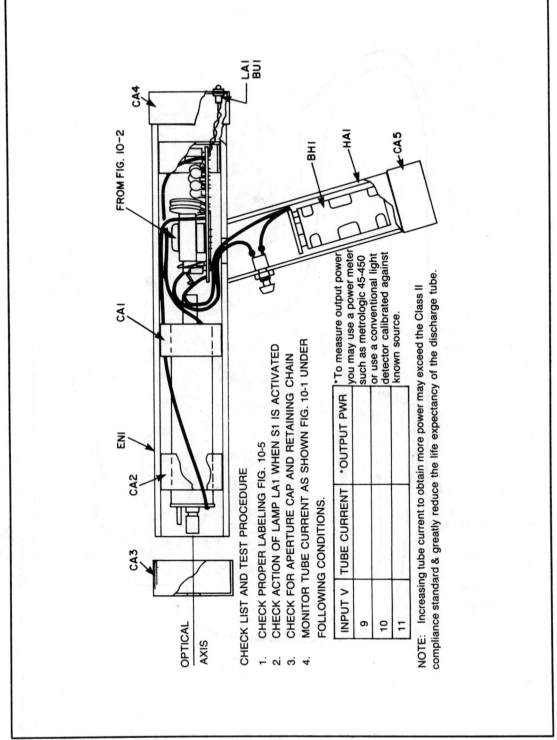

OPTICAL AXIS

CHECK LIST AND TEST PROCEDURE

1. CHECK PROPER LABELING FIG. 10-5
2. CHECK ACTION OF LAMP LA1 WHEN S1 IS ACTIVATED
3. CHECK FOR APERTURE CAP AND RETAINING CHAIN
4. MONITOR TUBE CURRENT AS SHOWN FIG. 10-1 UNDER FOLLOWING CONDITIONS.

INPUT V	TUBE CURRENT	*OUTPUT PWR
9		
10		
11		

*To measure output power you may use a power meter such as metrologic 45-450 or use a conventional light detector calibrated against known source.

NOTE: Increasing tube current to obtain more power may exceed the Class II compliance standard & greatly reduce the life expectancy of the discharge tube.

Fig. 10-3. X-ray view (1mW) LGU2/4.

122

of the power supply. Further tailoring of the laser tube parameters can be accomplished by inserting forward-biased silicon diodes (0.6 V) or germanium diodes (0.3 V). This is sometimes required to reduce excessive laser tube current. Changing the number of batteries may be too coarse of a voltage difference, providing too large a current change.

CONSTRUCTION STEPS

1. Layout and identify all parts and pieces by using Table 10-1. Recheck values and note components with polarity requirements but difference values.

2. Layout assembly board and wire as shown Figs. 10-1 and 10-2. Note proper position of transformer T1, and polarity of semiconductors. Drill small holes for parasitic diodes D5 to D10. Use the actual leads of the components for wiring whenever possible. Avoid bare wire bridges and verify all solder joints. Avoid sharp points and pay attention to component clearances in "parasitic multiplier" section.

3. Connect wire leads that attach to the actual laser tube. Do not connect to tube at this time. Connect short leads to LA1 visual indicator.

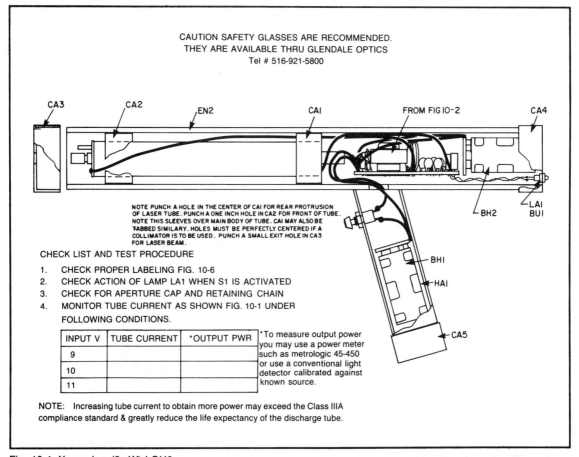

CAUTION SAFETY GLASSES ARE RECOMMENDED.
THEY ARE AVAILABLE THRU GLENDALE OPTICS
Tel # 516-921-5800

CA3 CA2 EN2 CAI FROM FIG 10-2 CA4

NOTE PUNCH A HOLE IN THE CENTER OF CAI FOR REAR PROTRUSION OF LASER TUBE. PUNCH A ONE INCH HOLE IN CA2 FOR FRONT OF TUBE. NOTE THIS SLEEVES OVER MAIN BODY OF TUBE. CAI MAY ALSO BE TABBED SIMILARLY. HOLES MUST BE PERFECTLY CENTERED IF A COLLIMATOR IS TO BE USED. PUNCH A SMALL EXIT HOLE IN CA3 FOR LASER BEAM.

BH2 LAI BUI

BHI

HAI

CA5

CHECK LIST AND TEST PROCEDURE

1. CHECK PROPER LABELING FIG. 10-6
2. CHECK ACTION OF LAMP LA1 WHEN S1 IS ACTIVATED
3. CHECK FOR APERTURE CAP AND RETAINING CHAIN
4. MONITOR TUBE CURRENT AS SHOWN FIG. 10-1 UNDER FOLLOWING CONDITIONS.

INPUT V	TUBE CURRENT	*OUTPUT PWR
9		
10		
11		

*To measure output power you may use a power meter such as metrologic 45-450 or use a conventional light detector calibrated against known source.

NOTE: Increasing tube current to obtain more power may exceed the Class IIIA compliance standard & greatly reduce the life expectancy of the discharge tube.

Fig. 10-4. X-ray view (2mW) LGU6.

4. Obtain a 9.8 to 10 volt source such as "8" AA fully charged nicad batteries. Connect a 220 kΩ 1-watt test resistor across the laser tube leads. Connect battery pack and measure a current draw of approx. 1 amp when meter is connected across the normally open contacts of switch S1. Reconnect the meter in series with one of the laser tube leads to the test resistor and measure 4 to 5 mA when S1 is pressed. Voltage across the test resistor should be 1100 volts. See Table 10-2 for load lines with 10 V and 15 V, respectively.

5. Check "parastic multiplier" by noting arc jumping about 1/16 of an inch where leads to tube are brought close together. Arc should sustain. Do not allow for more than several seconds to verify operation.

6. Fabricate main enclosure (EN1) and handle (HA1) as shown in Figs. 10-3 and 10-4. (Also, see Figs. 10-5 and 10-6.) Fabricate metal clamp (BK1) form a 1/2″ wide strip of short aluminum (Fig. 10-7). Note larger housing when building the 2 mW version.

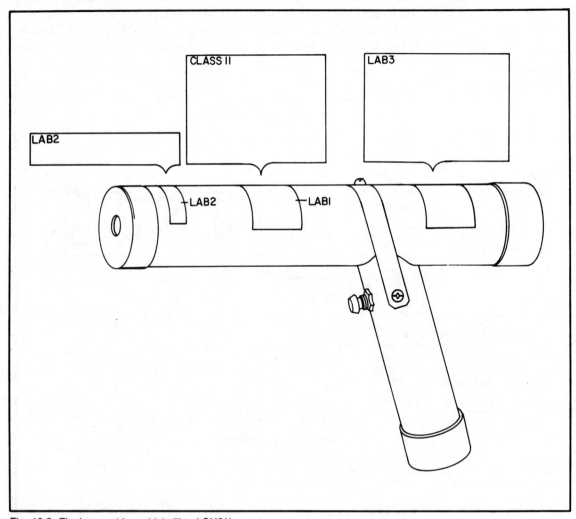

Fig. 10-5. Final assembly and labelling-LGU2/4.

124

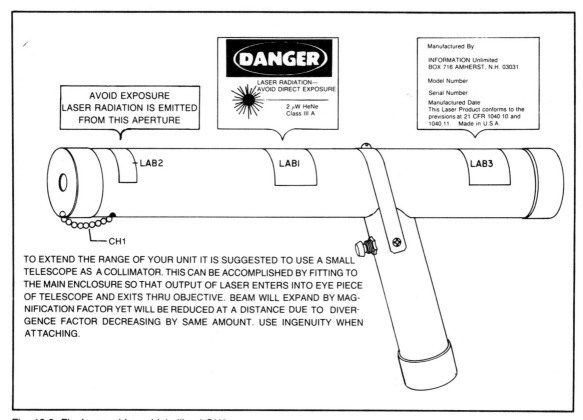

AVOID EXPOSURE
LASER RADIATION IS EMITTED
FROM THIS APERTURE

DANGER

LASER RADIATION—
AVOID DIRECT EXPOSURE

2 µW HeNe
Class III A

Manufactured By:

INFORMATION Unlimited
BOX 716 AMHERST, N.H. 03031

Model Number

Serial Number

Manufactured Date
This Laser Product conforms to the
previsions at 21 CFR 1040.10 and
1040.11.   Made in U.S.A.

LAB2

LAB1

LAB3

CH1

TO EXTEND THE RANGE OF YOUR UNIT IT IS SUGGESTED TO USE A SMALL
TELESCOPE AS A COLLIMATOR. THIS CAN BE ACCOMPLISHED BY FITTING TO
THE MAIN ENCLOSURE SO THAT OUTPUT OF LASER ENTERS INTO EYE PIECE
OF TELESCOPE AND EXITS THRU OBJECTIVE. BEAM WILL EXPAND BY MAG-
NIFICATION FACTOR YET WILL BE REDUCED AT A DISTANCE DUE TO DIVER-
GENCE FACTOR DECREASING BY SAME AMOUNT. USE INGENUITY WHEN
ATTACHING.

Fig. 10-6. Final assembly and labelling-LGU6.

7. Attach laser tube to power supply and verify series current of 4 to 5 mA with a freshly charged pack of batteries. Measure at ground or common return of tube. It may be necessary to use a diode dropper as noted in Fig. 10-1.

8. Fabricate laser tube "position bushings" (CA1, 2) by removing a 1″ wide circle or an amount equal to the diameter of the actual laser tube from the exact center of the plastic cap. Sleeve over tube and wire leads as shown Fig. 10-4.

9. Insert laser tube assembly into enclosure and position as shown in Fig. 10-5. Final assembly is as shown.

10. Label as shown in Fig. 10-5. Insert battery and caps CA3 and CA4. Note CA5 placed over output end to protect laser tube when not in use. A collimator may also be attached to increase the range of the system.

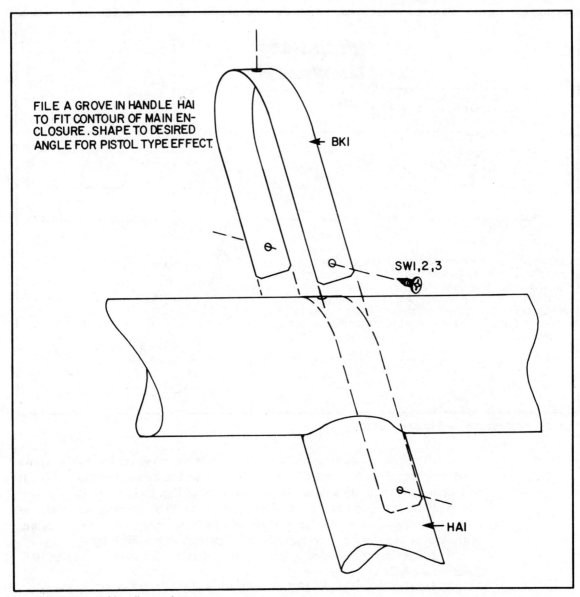

FILE A GROVE IN HANDLE HAI
TO FIT CONTOUR OF MAIN EN-
CLOSURE. SHAPE TO DESIRED
ANGLE FOR PISTOL TYPE EFFECT.

BKI

SWI,2,3

HAI

Fig. 10-7. Method of handle attachment.

Chapter 11

LASER RADIATION –
AVOID EYE OR SKIN EXPOSURE
TO DIRECT OR SCATTERED
RADIATION

3 JOULE RUBY

CLASS IV LASER PRODUCT

Ruby Laser Gun (RUB5)

Caution! Use of controls or adjustments or procedures other than those specified herein may result in hazardous radiation exposure.

ALL LIGHT, CONVENTIONAL OR LASER, IS THE RESULT OF EMISSION BETWEEN energy levels of atoms and molecules. These energy levels are characterized by certain quantum levels that are inherent to the particular atom. These particular levels have the property of resonance both in the absorption and emissions of radiant energy. This energy when emitted as radiation (light) is equal to E (joules) $= hf$ where h is Planck's constant and f is the frequency of radiation. Conversely, the frequency of the emitted radiation is functional to E higher $- E$ lower and applies to both laser and conventional systems.

Energy (radiation) is emitted (emission) when the atoms make a transition (change) from a higher level (excited state) to a lower level (relaxed or ground state). See Fig. 11-1. An atom, however, must first absorb energy (Fig. 11-2). All light requires a certain amount of absorption of energy by spontaneous emission and laser light is the result of stimulated emission. See the Electromagnetic Spectrum Sheet (Table 11-1).

It is obvious that spontaneous emission never produces light of the quality of laser light. Conventional light is the result of a system at thermal equilibrium where energy levels are always populated with the least in the highest levels and the most in the lowest levels. This condition always allows the atoms to be absorptive in nature only becoming saturated at infinite temperatures, (all energy levels filled). The basic character of conventional light is wide spectral distribution, random polarization, circular and irregular wave fronts, and relatively low resultant color temperatures.

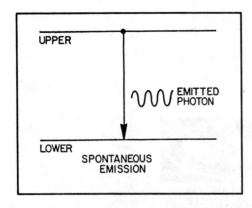

Fig. 11-1. The spontaneous emission of light.

A laser differs from conventional light in that the radiation produced is not by the spontaneous emission of energy levels, but by stimulated emission. This emission is the result of nearly equal levels of energy stimulating other levels into emitting photons of radiation of a nearly pure single frequency when relaxing to the lower energy states of the atom (Figs. 11-3 and 11-4). However, to obtain this stimulated emission effect, a *population inversion* consisting of now a maximum occupancy of the higher energy levels rather than the lower levels must occur and can only be achieved by forcing or pumping the system with external energy. This pumping or exciting is accomplished via flash lamps for pulsed solid-state devices, arc lamps for continuous solid-state, electrical discharge for gas systems, other lasers, or optically pumping an active medium, chemical reaction, etc.

Many materials will lase, however, only certain materials are worthwhile and produce usable output. Laser materials must be of an atomic structure where energy levels are able to be excited by practical means to achieve the necessary population inversion. Suitable materials are usually crystalline for solid devices and gaseous or liquid for others. These materials are usually placed between two mirrors where the radiation can make multiple passes through the materials stimulating more excited atoms into producing more radiation and so on until the radiation exits as a powerful beam of nearly pure light.

Of all the many laser devices and systems possible, the ruby laser takes precedence as being the first optical laser developed. (Schawlow, and Townes, early 1960). The ruby laser, while being extremely simple to construct, also produces one of the most

Fig. 11-2. The absorption of a photon.

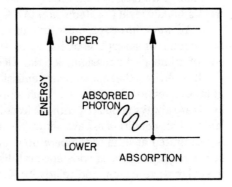

Table 11-1. The Electromagnetic Spectrum

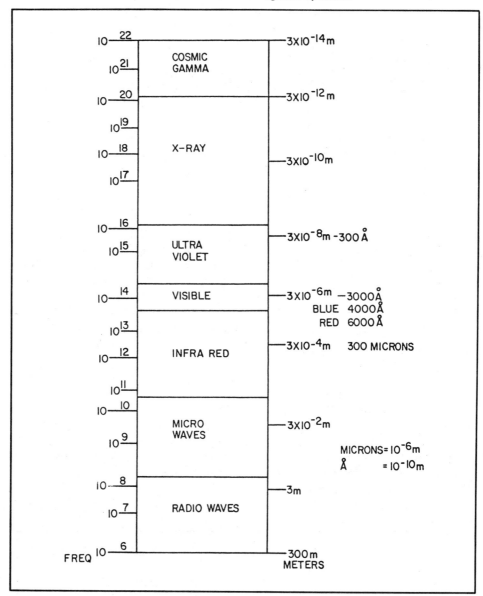

powerful sources of light when Q-switched. This type of device is referred to as a solid-state laser along with its neodymium-yag and neodymium-glass cousins.

The ruby laser, while being only fractions of a percent efficient (.1 percent), produces these high-powered optical pulses in the visible red range at 6943 °A. See Table 11-1. These pulses are capable of blasting holes through steel, bouncing off the moon, and there are special military systems capable of being used for antipersonnel weaponry when utilizing the more efficient neodymium-glass devices or the more recently discovered exotic mediums.

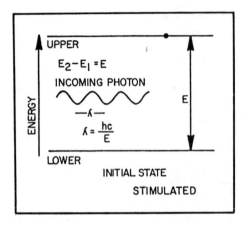

Fig. 11-3. The stimulated emission of a photon.

A ruby laser is a three-level device (Fig. 11-5) consisting of an active medium containing about 0.3 to 0.4% chromium CR^{+3} atoms to the main aluminum oxide (AL_2O_3) crystal host. The geometry of the medium is usually a cylinderical rod of from 1 cm to 2 cm in diameter and of 5 to 20 cm long. It is usually pink to red in color depending on the concentration of CR^{+3} atoms present in the host. The ends of the rod are usually optically coated with one being totally reflective and the other being partially reflective. The partial end is the output end. Light is reflected back and forth between these ends and is amplified by further stimulation of other atoms and eventually exits as a visible beam of laser light. Laser action is the result of the energy levels of the CR^{+3} atom.

Note that the 2E level is pumped with at least half the ions of the 4A ground state before laser action commences. The absorption of the ruby rod medium consists of two regions, one in the violet T1 and one in the green T2. These regions are about 1000 °A wide and are comparatively efficient when pumped by white light. Upon reaching the T1 and T2 absorption levels the excited ions quickly drop down to the 2E level creating the necessary population inversion required for laser action. The transition from this 2E level to the ground level produces the lasing output wavelength of 6943 °A. It is this transition from the 2E level to ground state that is the stimulated emission, i.e., atoms of nearly equal energy levels producing photons of energy that stimu-

Fig. 11-4. The stimulated emission of light.

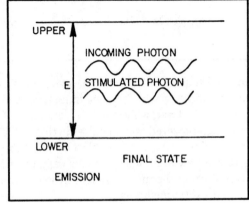

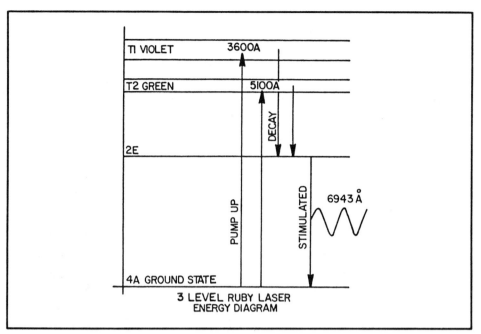

Fig. 11-5. 3 level ruby laser energy diagram.

late other atoms to produce more photons, etc. It should be noted that the diagram is greatly simplified and fails to show the closely separated defined lines of both the 2E and the 4A levels. These levels only serve to increase the spectral distribution of this output and can be "resolved" by cooling the rod to about 75 °K, when the line widths become about 10-15 GHz.

One now sees that the first step is to provide a pulse of light rich in green and violet to pump the ruby rod into the population inverted state. This is done via a xenon flash lamp that is physically placed near the ruby rod (Fig. 11-6).

There are several accepted methods of accomplishing the above, one being the use of a helical flash lamp placed around the ruby rod. This method is not quite as efficient

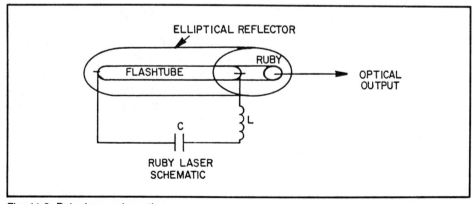

Fig. 11-6. Ruby laser schematic.

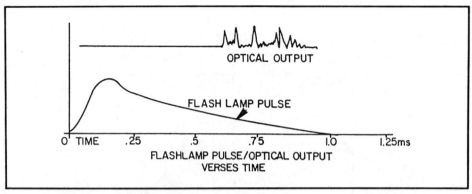

Fig. 11-7. Flashlamp pulse/optical output verses time.

as using linear flash lamps that are placed inside of a reflective cylinderical or elliptical reflector. The laser and lamps are located at their respective foci of these enclosures.

The laser pulse starts about 500 μsec after the pumping flash and lasts for another 500 μsec (Fig. 11-7). Depopulation of the upper levels occurs many times faster than the pumping rate can replenish them. Consequently, the system must rest until population inversion is again achieved. The output pulse consists of many spikes lasting for about 1 μsec each for each optical flash of the lamp.

The flash lamp of the system is usually energized by a conventional dc power supply charging a bank of low-loss capacitors. Voltages usually are from 1 to 4 kV with energy from 20 to 2000 joules. Joules = 1/2 CV. C = capacitance in farads, V = voltage in volts. This corresponds to about 50 to 1000 μF of capacity. You remember from high school physics that joules = energy = watts/sec. Therefore, watts = joules/sec. A typical moderately powered system may consist of the following: 500 μF at 2000 volts = 1000 joules. If the light pump pulse occurs in 5×10^{-4} seconds, the pulse power = 2 megawatts. A conversion efficiency of about 0.1% optical output will be a 2000 watt pulse lasting almost a millisecond. If, however, we were to Q-switch the above system at a pulse width of 10 nanoseconds, we would produce peak power of 1000 joules/10 sec = 100 GW optical power at an efficiency of 0.1% now produces an output pulse 1000 megawatts. Quite an impressive amount of optical intensity consisting of an equivalent color temperature when considering conventional sources at this bandwidth of astronomical proportions.

In order to obtain useful output from the ruby rod, it must be optically prepared to be totally reflective at one end and partially on the other (Fig. 11-8). This allows the optical energy to pass through the active medium of the rod on axis many time stimulating other transitions and increasing the (optical) energy on each successive pass and eventually exiting out of the partial end as the useful output beam. (This is termed a *Fabry-Perot* optical resonator). Q-switching is the spoiling of the above that allows a heavy population of the system creating a super high extremely short pulse of energy. Q-switching may be done mechanically, optically, etc.

BRIEF CIRCUIT THEORY

Your laser power supply is divided into two separate independent energy sources that in turn power one of the two flash lamps. See Fig. 11-9 (A) and (B). Energy storage is shown using four electrolytic capacitors being charged via a transformer and voltage doubler rectifier system. An ignition section supplies the HV trigger pulse to the flash lamps. A dual comparator circuit driving a power control relay prevents capacitor overcharge and allows independent voltage control via the two control pots. Voltage charge is maintained by the two meters on the control panel. Charge indication safety lamps are connected to the flash lamps in the laser head to help prevent accidental contact when the cover is removed. The capacitors hold a lethal charge for some time.

COMPLETE CIRCUIT THEORY

The project is described in four different sections respectively consisting of Basic Power Supply (Fig. 11-9); Ignitor, Delay and 12 V Power Supply (Fig. 11-10); Voltage Control Circuit (Fig. 11-11); and the Flash Head Section (Fig. 11-12) that is remotely connected via the umbilical cable. A convenient simplified control schematic Fig. 11-13 is also included along with a separate description of its functions described later in the chapter.

The Basic Power Supply, (Fig. 11-9) consists of two separate charging and storage sections that each supply one or the other flash lamps. High voltage is obtained by the step-up action of the transformer (T1, T2) through the appropriate control circuitry. Rectifier diodes (D1-D32) along with reverse voltage balancing resistors (R7-R38) rectify the high voltage and integrate this energy onto storage capacitors (C1 through C8). Power resistors (R5, R6) are adjusted for balancing the rate of charge between the two sections. They also can somewhat determine the actual charging time.

Resistor dividers R45 to R47 and R50 to R52 supply the current for the indicating meters (M1 and M2). Trimpots (R49, R53) are for meter calibration. A second set of resistor dividers (R54 to R61) supply the sensing current necessary for controlling the value of voltage charge on the energy capacitors. You will note pulse forming inductors (L1, L2) connected in series with the discharge lines to the flash tubes. These inductors are discussed in the section on flash tubes as to their exact function and value.

The Ignitor, Delay and 12-V Power Supply (Fig. 11-10) supply the high voltage trigger pulse to the flash lamp. This pulse is delayed by the charging action of (C2B) through resistor (R6B). Low voltage power is obtained from transformer (T3B) rectifier diodes (D9B to D12B) and filter capacitor (C4B). A power inverter consisting of

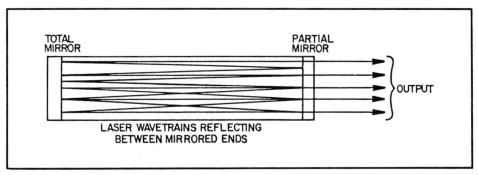

Fig. 11-8. Laser wavetrains reflecting between mirrored ends.

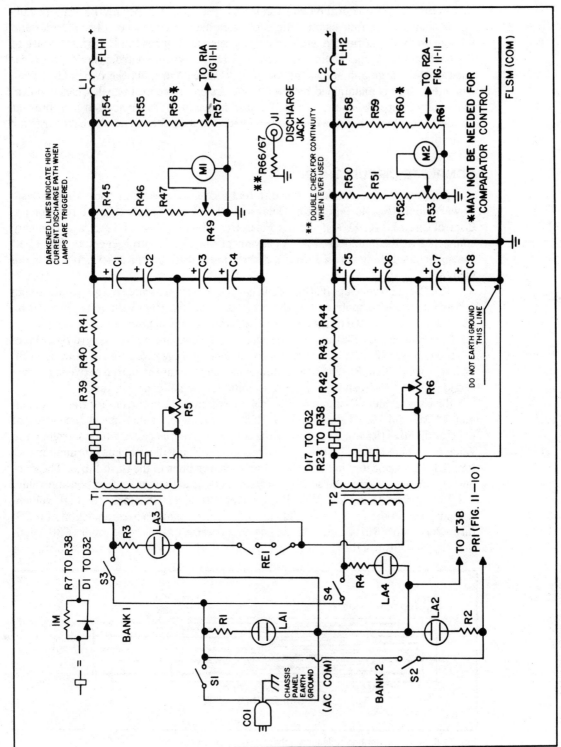

Fig. 11-9A. Ruby laser basic power supply.

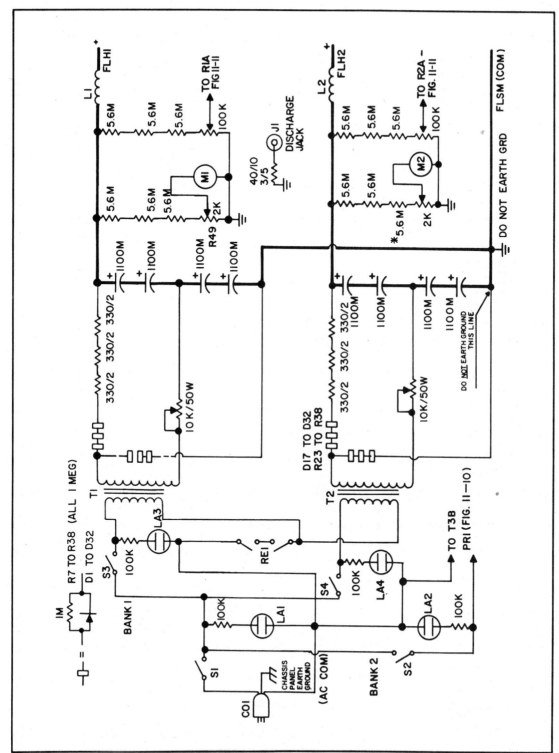

Fig. 11-9B. Ruby laser basic power supply basic component values.

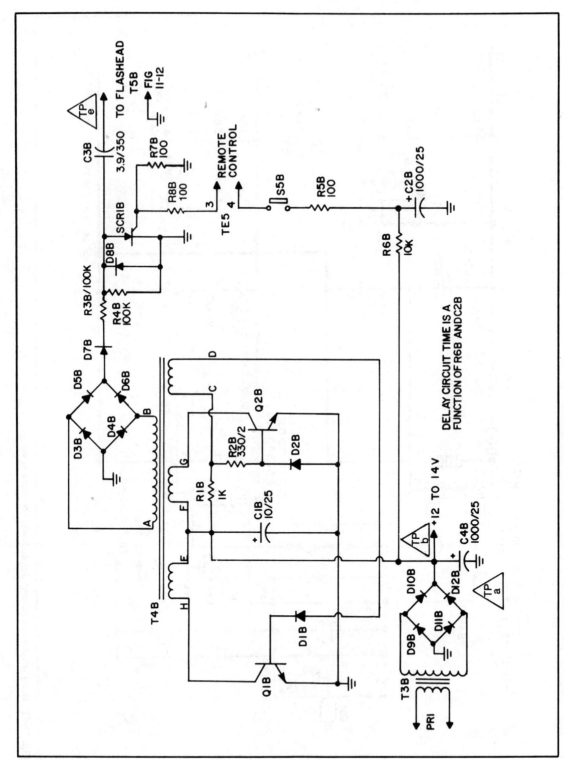

Fig. 11-10. Ignitor and delay/12 volt power supply; note component values shown.

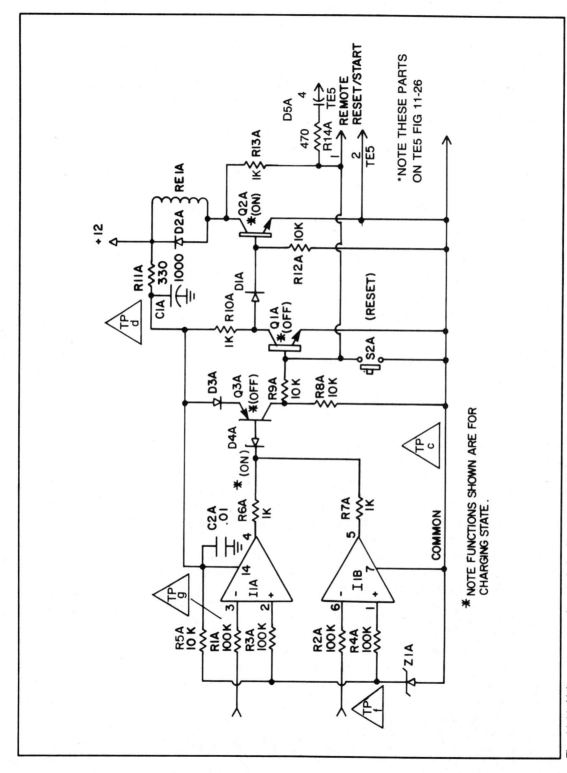

Fig. 11-11. Voltage sense and control circuit; note component values shown.

137

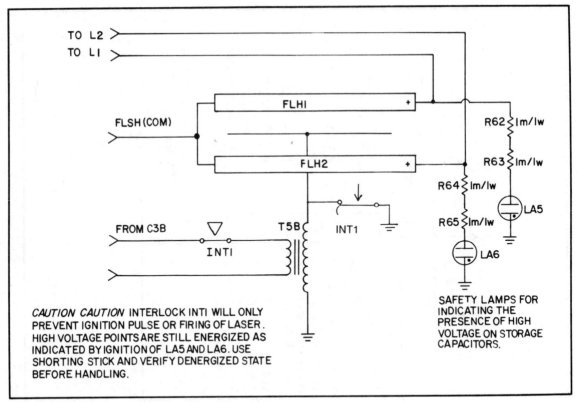

Fig. 11-12. Flash head schematic.

switching transistor (Q1B, Q2B) and saturating transformer (T4B) step 12-Vdc to 400-Vdc necessary for the capacitor discharge action of the circuit. T4B is a saturating transformer whose primary windings are switched by Q1B and Q2B. Diodes (D1B, D2B) provide the base return paths for the feedback current to their respective bases. Resistor (R2B) limits the base current while resistor (R1B) provides the necessary electrical imbalance to sustain oscillation. The output winding of T4B is rectified by diodes (D3B to D6B). This energy is integrated onto C3B through the primary of T5B.

Silicon-controlled rectifier switch (SCR1 B) upon actuation via its gate now dumps the energy from C3B into T5B primary, force inducing the high-voltage trigger pulse. Diode (D8B) recovers the overshoot ringing of the system adding to the circuit efficiency. Voltage dividing resistors (R3B, R4B) divide the dc charge voltage accumulated onto C3B to one half. The SCR is triggered by a positive pulse at its gate electrode. This is produced upon activation of switch S5B when C2B charges up.

In the Voltage Controller (Figs. 11-9 and 11-11) the sense voltage is obtained from R54, R55 and R58, R59 respectively for each bank. Control pots (R57) and (R61) set the level for the two voltage comparators (I1-A) and (I1B). Reference voltage is obtained through R5A and regulated to 5.2 volts by a zener diode (Z1A). The output of the comparators remains at a high until the sense voltage reaches the same value on pins 3 and 6 that the reference voltage set on pins 2 and 1. When this happens the outputs respectively on pins 4 and 5 go to zero and causes Q3A to conduct, biasing transistors (Q1A) "on." The collector of Q1A now goes to zero turning off relay driver transistor (Q2A). Indicator LED (D1A) now goes off showing that RE1A is now deenergized therefore stopping the charging of the capacitor banks. You will note that when switch S1A is open that RE1A will go "on and off" always "topping off" the charge on the capacitors. This may be undesirable as it creates transients and noise and should only be used in making adjustment to the voltage control pots R57 and R61.

In actual operation switch S1A would be closed preventing the system from charging upon "powering up" since relay RE1A cannot energize until S2A grounds the base of Q1A. This now allows RE1A to energize, thus recommencing the charge cycle. Now when the sense level from pots R57 and R61 reach the reference level RE1A locks off and remains so until S2A is again reset. It is now where the ignition will ignite the flash tubes via switch (S5B) thus discharging the capacitor bank. Recharging is now simply initiated by reactivating S2A. It should be noted that a power failure will require this reset action and is necessary to comply with the new BRH Class IV laser compliances. The remote control function consists of this reset function terminals 1 and 2 along with the gate control of the ignition terminals 3 and 4.

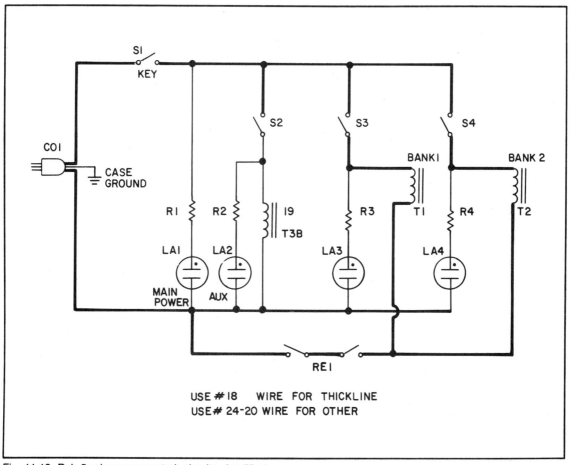

Fig. 11-13. Rub 5 primary ac control; circuits simplified.

FLASH LAMP AND ASSOCIATED PARAMETERS

Your power supply design specifications are chosen as a result of several inherent parameters required by the laser system. It is assumed that a flash lamp filled with xenon will be used as this is the most efficient gas for this purpose while still supplying the proper spectral output either for yag, neodymium or ruby lasing mediums.

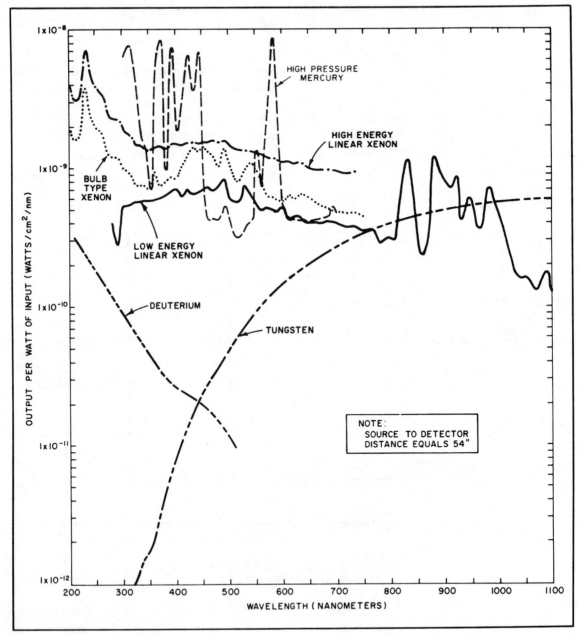

Fig. 11-14. Xenon flashlamp spectrum versus other commonly used optical sources (courtesy of EG&G Electro Optics).

It is a known fact that a ruby crystal favors the green and purple part of the spectrum while yag and neodymium favor the infrared region. This spectral shift can be optimized by applying a different ratio of charging capacitor value to charge voltage keeping the charge energy in joules constant. Color temperature varies between 5000 to 10,000 °K. This spectral shift is due to the current density through the flash tube at the time of discharge (higher current favoring the lower wavelengths with lower current being vice versa. This is shown in Fig. 11-14.

Another parameter is the energy required to threshold the lasing medium (Ruby requiring higher energy than yag). We assume that our rod will require 150 joules per inch for a total of 450 joules. Flash lamp efficiency is usually around 50%, therefore requiring 900 joules to the two lamps. Since we are using two 3 inch lamps the value becomes 150 joules per inch of flash lamp length.

Our next parameter is flash pulse duration. This usually runs 500 to 1000/μsec and is determined by the inherent physics of the lasing medium. We have chosen 500/μsec for the system we describe, as most ruby crystals appear to accept this value.

An important consideration at this point is the anticipated flash repetition rate we require to determine the cooling requirements of the system. We now can calculate the average power we need to dissipate by the simple formula (watts) = flash rate (pulse per sec) × pulse energy (joules). Our system is intended for free air at a repetition rate of one pulse every 3 minutes. (Note that forced air can be added as an option and allows an increase in repetition rate). We now refer to Table 11-2 for a convectively cooled end cap type flash tube showing 25 watt/cm (where cm is the inner wall area of the lamp). This equates to 450 joules × 1/30 = 15 watts. The inside area of the tube wall we have chosen is 14 cm. Therefore, we have a 14 × 25 = 350 watt rating being well below the value of 15 required.

Table 11-2. Power Limitations of Flashlamps with Various Seals and Cooling Methods (courtesy of EG&G Electro Optics).

Type of Seal	Method of Cooling	Maximum Wall Loading (watts/cm^2)
End-Cap	Convection	25
End-Cap	Forced air	50
End-Cap	Water	350
Graded	Convection	25
Graded	Forced air	50
Graded	Water	300
Butt	Convection	25
Butt	Forced air	50
Butt	Water	Not recommended

Our next problem is to determine a proper lamp at the above size that will give us the above parameters and yet last for hundreds of shots and not be in danger of exploding. This is where we consult a manufactures specification guide in this case the EG&G Flash Lamp Manual.

Note Fig. 11-15 and use the above determined parameters to help select an actual tube. Remembering we have determined 150 joules per inch for 500/μsec, we now check our available options. Our approach utilized a 5mm ID × 6mm OD that allows for over 200 joules per inch. We therefore choose EG&G tube number FX103C. This tube is shown in Table 11-3 with other tubes. Specifications are divided into electrical and mechanical.

The following formulas are for highest efficiences utilizing linear flash lamps when considering dampening factors.

The next step is to determine the electrical parameters necessary to fit those already obtained. The storage capacitor is our next objective and is the part that stores the necessary energy that is used to supply the actual flash. The value is determined by:

C = microfarads/lamp

E = joules = 450

α = Damping factor (.8 for critical) See Fig. 11-16.

T = t/3 pulse duration = 166 μsec

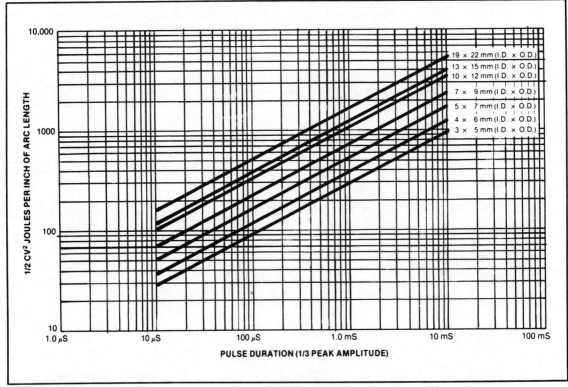

Fig. 11-15. Explosion energy for various flashlamp quartz bore sizes (courtesy of EG&G Electro Optics).

TUBE DESCRIPTION	FX-147C-2	FX-IC-6	FX-5C-9	FX-33C-1.5 FX-84C-1.5	FX-33C-2 FX-38C-2 FX-103C-2	FX-38C-3 FX-85C-3 FX-103C-3	FX-98C-3	FX-42C-3	FX-45C-6 FX-52C-3	FX-81C-4 FX-55C-6	FX-81C-6.5	FX81C-8
Quartz Envelope (ID × OD mm)	3 × 5	4 × 6	4 × 6	4 × 6	4 × 6	4 × 6	5 × 7	7 × 9	7 × 9	10 × 12	10 × 12	10 × 12
ARC Length (inches/millimeters)	2/51	6/152	9/229	1.5/38	2/51	3/76	3/76	3/76	6/152	4/102	6.5/165	8/203
ELECTRICAL CHARACTERISTICS												
Minimum Starting Voltage, V (kVdc)	0.5	0.7	1.2	0.5	0.6	0.7	0.8	1.0	1.1	1.0	1.0	1.0
Minimum Trigger Voltage (1) V_T (kV)	15	20	20	15	15	15	15	20	20	25	25	25
AT τ = 100 μs FLASH DURATION (2)												
Maximum Energy Input, E (3) (joules)	133	500	756	126	168	252	336	462	924	980	1592	1960
Voltage, V (kVdc)	2.1	5.3	7.9	1.4	1.8	2.6	2.5	2.1	2.4	2.4	3.9	4.9
Inductance, L(μH)	28	45	68	13	15	23	15	8	15	5	8	10
Capacitance, C(μF) (5)	58	36	24	128	106	71	106	212	106	320	212	160
Tube Impedance, R_T (Ω) (5)	1.1	1.8	2.7	0.5	0.6	0.9	0.6	0.3	0.6	0.2	0.3	0.4
AT τ = 1000 μs FLASH DURATION (2)												
Maximum Energy Input, E (3) (joules)	420	1680	2520	420	560	840	1050	1490	2980	3080	5000	6160
Voltage, V (kVdc)	1.35	2.8	5.3	0.89	1.2	1.8	1.6	1.36	2.7	1.7	2.5	3.1
Inductance, L(μH)	350	600	900	150	200	300	200	100	200	75	100	125
Capacitance, C(μF) (5)	460	265	177	107	800	535	800	1600	800	2120	1600	1280
Tube Impedance, R_T (Ω) (5)	1.4	2.4	3.6	0.6	0.8	1.2	0.8	0.4	0.8	0.3	0.4	0.5
MAXIMUM AVERAGE POWER (watts) (4)												
Convection (25 °C Ambient)	15	40	50	10	15	20	30	60	80	80	90	100
Forced Air (10-15 cu ft/min)	120	320	400	80	120	160	240	480	640	640	720	800
Water (1 gpm at 25 °C)	900	2400	3000	500	900	1000	1800	4000	8000	8000	9000	10,000
LIFE IN NUMBER OF FLASHES	See Life Chart Figure 2											

New EG&G Part Number	Old EG&G Part Number	K_o Impedance (ohm-amp$^{1/2}$)	K_{ex} Explosion Energy joules (μs$^{1/2}$)	Maximum Average Power			Maximum Peak Current (amps)	Operating Voltage		Trigger Pulse	
				Convection Cooled	Forced Air Cooled	Water Cooled		Min.	Max.	kV	$t_{1/3}$ μs
4XCG-1.5	FXG-33C-1.5	12.2	18	10	80	500	500	500	1600	15	0.3
4XC-3	FX-38C-3	24.3	36	20	160	1,000	500	700	2500	15	0.6
4XC-3A	FX-103C-3		36	20	160	1,000	500	700	2500	15	0.6
7XC-3	FX-42C-3	13.9	67	60	480	4,000	1400	1000	3000	20	0.6
7XC-3A	FX-227C-3	13.9	67	60	480	4,000	1400	1000	3000	20	0.6
7XCG-19.5	FXG-160C-19.5	90.5	440	Consult Factory			1400	1500	7300	25	3.9
13XC-6.5	FX-47C-6.5	16.2	270	100	800	10,000	2800	1000	3300	20	1.3
13XC-6.5A	FX-397C-6.5	16.2	270	100	800	10,000	2800	1000	3300	20	1.3
13XC-6.5B	FX-469C-6.5	16.2	270	100	800	10,000	2800	1000	3300	20	1.3

$$K_o = \text{Impedance of lamp} = 24.3 \text{ (Table 11-4)}$$

$$*C = [2E_o \alpha^4 T^2 K_o^{-4}]^{1/3} \cong (900)(.41)(27.6)(10^2(2.9)(10^{-6})(1/3)(29534)(10^{-15})(1/3) \cong 30.9 \times 10^{-5} \cong 309 \; \mu F$$

*The damping factor is very important in power supply design for flash lamps. It determines such effects as (negative overshoot) being the result of ringing. Peak current through the tube is a function of this damping factor being less for values over 0.8 and more for values under 0.8. Under damping also results in the undesirable negative overshoot that reduces flash tube life. Several curves are shown in Fig. 11-16. Damping values increase with inductance and decrease with resistance. Critical damping is 0.8 for optimum results.

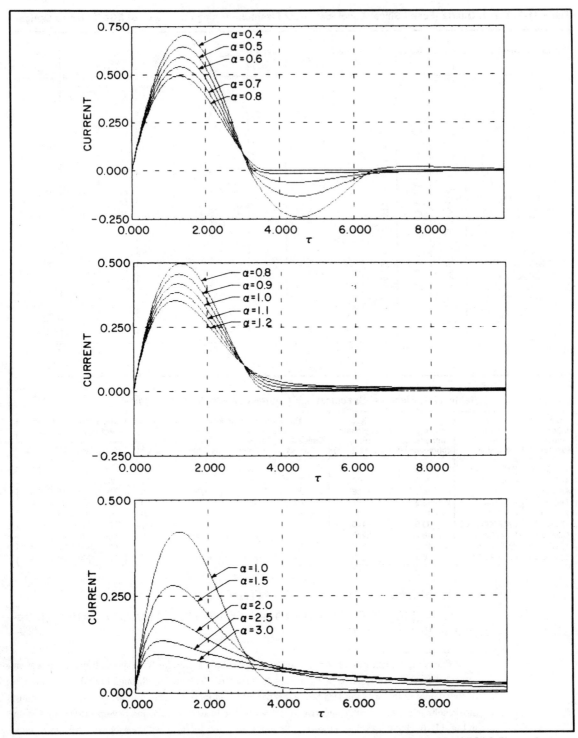

Fig. 11-16. Normalized current, I, as a function of normalized time, τ, for $\alpha = 0.4$ to 3.0 (courtesy of EG&G Electro Optics).

We now determine the inductance L required from the formula:

$$*L = \frac{T^2}{C} \cong \frac{(27.6)(10^3)(10^{-12})}{(309)(10^{-6})} \cong 70 \ \mu H$$

We now determine the capacitor charging voltage V required from the formula:

$$*V = \left(\frac{2E_o}{C}\right)^{1/2} \cong \frac{900}{(309)(10^{-6})} \cong 1.7 \ kV$$

The last computation involves determining the parameter for our inductor regarding turns, etc. for the value of 100 μH. We select a coil form of 1.5 inches in diameter determined by prior experience and use the standard air coil formula.

$$L(\mu H) = \frac{a^2 n^2}{9a + 10b}$$

A final check using the formula $T = LC$: $t = 3T = 3 \ LC = 500 \ \mu sec$ (that verifies our pulse duration).

LASER RODS

This is a special note on performance expected using certain types of laser rods. LAB2 must state wavelength and energy.

Ruby. This rod will produce visible red pulses that are the least efficient. Its advantage is it can be "Q" switched where extremely high-powered short pulses are possible. It is shown in the plans. Q-switching is not discussed as it is usually beyond the homebuilder's facilities.

Yag. This rod is the most expensive and the most efficient. It produces 1.06 μm pulses of 5 to 10 times the efficiency of the ruby rod.

Neodymium-Glass. This rod is the most practical for this system. It is available "cavity ready" and can produce high-powered low-repetition-rate pulses at 1.06 μm.

Optional mirror mounts are shown for using rods without the integrated mirrors. However, we recommend the fully prepared neodymium-glass as this approach eliminates much hassle in optical alignment.

Your system is shown using the visible red ruby rod. This was the laser that pioneered the field. The rod is made by first growing a crystal boule formed from molten aluminum oxide (AL_2O_3) and doped with the proper concentration of chromium (Cr). There are several methods of producing these boules and all processes may be used depending on cost and optical quality desired. The boule now is machined into

---

*These parameters are not exact but can certainly be used as a starting point in final selection of these values. As an example the available capacitor values used are (4) 1100 $- \ \mu F$ 550-volt electrolytics in series. These equate out to 275 μF at a rating of 2000 volts. We now fudge the above equation to compensate for this value and come out with a value of 1810 volts to make up our 450 joules of energy. Also we have moved our parameter to favor the lower wavelength part of the spectrum necessary for ruby excitation. Note that we have changed our damping factor to 0.73 and inductance to 100 μH.

the required sized rod. The ends are now either antireflective coated or may have integral dielectric mirrors eliminating the need for outside optics. The system shown is intended for use with a laser rod with the integral mirrors coated on the ends. Optical mirror mounts are shown for those desiring the external mirror approach.

Rods without the integral mirrors require properly positioned external optics with a mechanically stable system for proper maintenance of their adjustment. An advantage to external mirrors is that damage is less frequent, therefore they are considered more for higher energy systems.

Performance of your system will depend on the type and quality of the rod used. Do not expect to obtain the beam quality of gas lasers. There is no way we can dictate the exact parameters that you may obtain from your system. The flash lamp energy at 1500 volts is 600 joules and using a conservative rating of flash lamp to output energy it is possible to obtain up to a joule of optical energy output.

Special note: Some of our customers have reported that helium-neon laser mirrors will work for lower powered systems using the ruby laser rod. We have not verified this in our labs at the time of this writing.

Special note: The laser head may be built to accommodate most 2-inch to 4-inch rods. A 2-inch rod may be used with correspondingly shorter lamps. Output will be reduced.

CONSTRUCTION STEPS

1. Start by laying out and identifying all the components for the assembly boards as shown in the appropriate figures and tables. Follow the layout as shown in the pictorials and use the leads of the components for the connecting points whenever possible. Avoid wire bridges and short circuits. Allow some lead length on components for heatsinking when soldering. Note the polarity of certain components and leave off interconnecting wires to other locations at this time. See Tables 11-5 through 11-10.

2. Assemble the "rectifier and resistor" board as shown in Fig. 11-9. Figure 11-17 shows the layout and routing of the wiring. It is suggested to follow the layout as shown as there are high voltage points that must be properly spaced. Interconnecting wires may be attached during final wiring. Note the screw locations for securing the finished board in place. Note the polarity of the 1N4007 diodes and leave adequate service leads for replacing, heatsinking, etc. Each diode has a 1 MΩ resistor in parallel with it for proper balancing of any peak reverse or transient voltages.

3. Assemble the "igniter/12-Vdc power board" as shown by schematic Fig. 11-10 and layout drawing Fig. 11-18. Wire as shown using the leads of the actual components for connections. Avoid wire bridges and attempt to follow the layout as shown. Note polarity of diodes, transistors, and capacitors. Do not attach interconnecting wires at this time. Transformer T4B can be identified by measuring the higher resistance between pins A and B before inserting.

4. Assemble the "sense and control board" as shown by schematic Fig. 11-11 and layout drawing Fig. 11-19. Wire as shown using the leads of the actual components for connections. Avoid wire bridges and attempt to follow the layout as shown. Note polarity of diodes, transistors, and capacitors. Do not attach interconnecting wires at this time. Note the coil of RE1A relay is connected via dashed lines.

Table 11-5. Power Supply Basic Parts List/Flash Head.

Ref.	Qty.	Description
R1,2,3,4	4	100 kΩ 1/4-watt resistor
R5,6	2	10 kΩ 50-watt adj resistor
R7-R38	32	1 MΩ 1/4-watt resistor
R39-R44	6	330 Ω 2-watt resistor
R45,46,47		
R50,51,52		
54,55,56		
58,59,60	12	5.6 MΩ 1-watt resistor
R49,53	2	2 kΩ trimpot resistor
R57,61	2	100 kΩ pot resistor
R62-65	4	1 MΩ 1-watt resistor on laser head
R67	1	10 kΩ 50-watts
R66	1	40-Ω 10-watts
*C1-8	8	1100 µF 450 V caps (See text)
D1-32	32	1N4007 1000 V lamp diodes
LA1-LA6	6	Neon lamp with leads LA5,6 on laser head
*S1		Key switch (nonremovable key in "on")
S2,3,4,1A	4	SPST 5-amp switch
*T1,2	2	1200 V 200-mA transformers
*L1,2	2	Inductors (See text) 45 turns #12 magnet wire
M1,2	2	0-100 µa meter
CO1	1	Line cord 3 wire #16
PB1	1	9″ × 4 3/4″ .1″ × .1″ grid perforated board stock
TE1,2,3,4	4	7-lug barrier strip
TE5	1	4-screw terminal strip
*T5	1	45 kV ignitor pulse transformer on laser head
J1	1	Jack

Mechanical Parts Main Supply

Ref.	Qty.	Description
FP1	1	Front panel and side A fab from Fig. 11-22
RP1	1	Rear panel and side B fab from Fig. 11-21
PM1,2	2	Plywood base mount top and bottom
COVER	1	Fab to cover components and protect from accidental contact. Use perforated metal, screen, etc.
TY1	12	Heavy wraps 12″
CL1	6	Nylon clamps, 1″, 1/2″, 1/4″ for routing and securing wires
WN1	2	Hi3 wire nuts
WN2	12	Hi0 small wire nuts (unused leads from transformer)
BU1	1	Linecord bushing
BU2	1	Umbilical bushing 1/2″
HDWRE		Assorted screws, nuts, washers and miscellaneous hardware for assembly
LAB1	7	HV warning labels
LAB2	1	Class IV compliance labels
LAB3	1	Certification label

147

Table 11-6. Ignitor 12-Volt Power Supply Parts List.

R1B	1	1 kΩ 1/4-watt resistor
R2B	1	330 Ω 2-watt resistor
R3B,4B	2	100 kΩ 1/4-watt resistor
R6B	1	10 kΩ 1/4-watt resistor
R5B,7B	2	100 Ω 1/4-watt resistor
C1B	1	10 µF/50-V elect
C2B,4B	2	1000 µF/50-V elect
C3B	1	3.9 µF/350-V
D1B,2B,9B 10B,11B, 12B	6	1N4001 50-V 1-A rectifier
D3B to D8B		1N4007 1000-V 1-A rectifier
*Q1B,Q2B		D40D5 NPN power tab
*SCR1B	1	C107D SCR
*T4B	1	Type 1 PC ferrite transformer
S5B	1	Pushbutton switch
T3B	1	12-V 300-mA/115-V transformer
PB2	1	6″ × 2 1/2″ .1″ × .1″ grid perforated board stock

5. Fabricate rear and front panel section RP1 and FP1 respectively as shown in Fig. 11-23 from a 28-1/2 inch by 4 inch piece of #24 galvanized stock. Note that this section is also the sides and contains an upper and lower 1/2″ lip for attaching to the wooden bases. Layout rear panel section as shown in Fig. 11-21 and Fig. 11-22 for front panel sections. Follow layout and drill holes to match size of components.

6. Fabricate (PM1) plywood mounting base from a 14 inch by 14 inch piece of 3/4 inch finish stock. Note the passage holes for the interconnecting wires that go to the storage capacitors (C1 through C8) and pulse shaping inductors (L1 and L2). Trans-

Table 11-7. Voltage Control Parts List.

R1A,2A 3A,4A	4	100 kΩ 1/4-watt resistor
R5A,8A 9A,12A	4	10 kΩ 1/4-watt resistor
R6A,7A 10A,13A	4	1 kΩ 1/4-watt resistor
R11A	1	330-Ω 2-watt resistor
R14A	1	470-Ω 1/4-watt (mounted on TE5)
C1A	1	1000 µF 20-V electrolytic
C2A	1	.01 µF/50-V disc
D1A	1	LED 10 mA green
D2A, D5A	2	1N4007 1000-V 1-A rectifier (D5A mounted on TE5)
D3A,4A	2	1N914 small signal
Z1A	1	6.2-V zener 1N5234
Q1A	1	PN2222 NPN GP transistor
Q2A	1	D40D5 NPN pwr tab transistor
Q3A	1	PN2907 PNP plastic transistor
I1A,B	1	LM3900 4-quad IC 14-pin DIP
RE1A	1	5-amp relay/12-V coil
S2A	1	Push button (no)
PB3	1	3″ × 4″ .1″ × .1″ grid perforated board stock

Table 11-8. PL Laser Head (Mech).

*CAV1	1	2 part cavity 3″ double ellipse silvered
*BLC1,2	2	Teflon MTG blocks 1-1/4″ × 1″ × 3/4″ (see text)
*SPC1,2	2	Teflon spacers cavity 3/8″ × 3/8″ × 3″ (see text Fig. 11-37)
*HOLD 1, 2,3,4	4	Teflon rod holders (see text Fig. 11-37)
CLAMP	1	For holding cavity together use 2″ wide metal strip fab per Fig. 11-35 inset
BRKT	1	Bracket for connection ignitor and retaining rod holders use 1″ wide metal strip per Fig. 11-36 inset
INTL	1	Interlock switch fab from copper (see text Fig. 11-39)
ADJ1	2	Mirror adj plate fab per text (Fig. 11-41)
REAR	1	Rear mirror mount fab per text (Fig. 11-42)
EXIT	1	Exit mirror mount fab per text (Fig. 11-41)
EPL1	2	Exit panel fab per (Fig. 11-33)
END	1	Rear panel fab see text (Fig. 11-34)
ROD1	2	1/4-20 rods threaded/20 nuts and washers
SPRING	6	Small compressor springs
FCL1	4	Flash tube clips
CAPS	2	Brass protective caps for laser mirrors (see text)
WR1	1	6″ HV ignition wire
BU1	1	Clamp bushing for umbilical wires
BU2,3,4 5	4	3/8″ plastic bushings
TUB1	1	2″ PVC 2″ × 1 5/8″ tubing (Fig. 11-38)
TUB2	1	6″ PVC 6″ × 1 1/2″ tubing (Fig. 11-38)
LRT1	1	Lens retainer see text (Fig. 11-38)
LEN1	1	Lens 29 × 43
AP1	1	Aperture cap and chain (Fig. 11-38)
R62-65	4	1-MΩ 1-watt
NEONS	2	Neon lamps
BPL1	1	Bottom base plate (fab per Fig. 11-32 A and B)
SHIM		Shim stock use PC stock (see text)
ACT1	1	Interlock actuator (Fig. 11-40)
PRH1	1	Protective housing fab see text (Fig. 11-40)
HDWRE		Misc screws, nuts and hardware for assembly
LAB4	1	Aperture label type E (Fig. 11-38)
LAB5	1	Interlock label type N (Fig. 11-40)

formers (T1 and T2) require holes for mounting screws as well as clearance holes for the primary and secondary wires. **Caution! These transformers must be secured via actual screws, nuts, and washers—do not use wood screws.** Note plastic clamps for securing the inductors in place. The capacitors are secured via heavy

Table 11-9. Switch Function Chart.

S1	Key switch enables entire system to be energized.
S2	SPST toggle switch powers up low-voltage circuits for ignition and sensing circuits.
S3	SPST toggle energizes capacitor Bank I.
S4	SPST toggle energizes capacitor Bank II.
S5B	Pushbutton triggers laser action via ignition to flash tubes—also in conjunction with remote control.
	All the above parts available from Information Unlimited. P.O. Box 716, Amherst, NH 03031
	Parts with asterisk are available individually.
S2A	Pushbutton is required for normal operation when S1A is closed. Initiate charge cycle. Also resets system for BRH class IV compliances.

Table 11-10. Wire Charts to Be Used as A Guide.

The following is a list of the wires used in the system. They are divided into the '' wiring aid sections.''

Wires Referenced Fig. 11-26

From R41 to TE2 #3 (use #20 HV wire at least 3 kV)
From R44 to TE2 #5 (use #20 HV wire at least 3 kV)
From R6 to C6, C7 (use #20 HV wire at least 3 kV)
From R5 to C2, C3 (use #20 HV wire at least 3 kV)
From TE2 #3 to C1+ (use #20 HV wire at least 3 kV)
From TE2 #5 to C5+ (use #20 HV wire at least 3 kV)
From L1 to TE3 #3 (use lead of L1, dress for clearance)
From L2 to TE3 #5 (use lead of L2, dress for clearance)
*From L1 splice to #14 wire with wire nut, to laser head
*From L2 splice to #14 wire with wire nut, to laser head
From TE2 #7, #14 wire to C4, C8 common
*From TE2#7, #14 wire to laser head
From TE4 #4 #20 wire to C4, C8 common
*These are umblical lead to laser head-must be flexible and no longer than 10 feet to comply with the BRH for deletion of emission indicator in laser head
From J1 to TE4 #7 (use #20 wire)
From C8 negative to C4 negative (use #14 wire with button hook).

Wires Referenced Fig. 11-27

From T1 to TE1 #1 & 2 (use red wires from T1)
From T2 to TE1 #3 & 4 (use red wires from T2)
From R6 to TE1 #3 (use #20 HV wire)
From R5 to TE1 #1 (use #20 HV wire)
From TE1 #2 to diode mid-point Bank I (use #20 HV wire)
From TE1#4 to diode mid-point Bank II (use #20 HV wire)
Dead end and wire nut all unused wires from T1 and T2
From arms to R5 and R6 ends (use #20 wire)

Wires Referenced Fig. 11-28

From TE1 #5 to RE1A (use #18-20 600-V vinyl hook up wire)
From TE1 #6 to S3 (use #18-20 600-V vinyl hook up wire)
From TE1 #7 to S4 (use #18-20 600-V vinyl hook up wire)
From TE3 #4 to S2 (use #18-20 600-V vinyl hook up wire)
From TE3 #6 to RE1A (use #18-20 600-V vinyl hook up wire)
From RE1A to lamp common and lamp to lamp (use #24 vinyl hook up wire)
From S1 to S2 to S3 to S4 (use short pieces #20 hook-up wire)
Use actual leads of T1, T2 and T3B as shown to TE1 and TE3 respectively
Power cord CO1 connected as shown

Wires Referenced Fig. 11-29

From S5B to TE5 #4 #24 hook up
From S5B to R5B on Fig. 11-18
From S2A to S1A short jump #24 hook-up wire
From S1A to Base Q1A #24 hook-up wire on Fig. 11-19
From S1A to R13A #24 hook-up wire on Fig. 11-19
From S2A to common line #24 hook-up wire on Fig. 11-19
From TE5 #1 to Bas Q1A #24 hook-up wire on Fig. 11-19
From TE5 #2 to common line hook-up wire on Fig. 11-19

Wires Referenced Fig. 11-30

From TE3 #2 to C3B (use #20 hook-up wire) on Fig. 11-18
From TE3 #3 to common (use #20 hook-up wire) on Fig. 11-18
From TE3 #2 to laser head (use #20 hook-up wire)
*From TE3 #3 to laser head (use #20 hook-up wire) Umbilical cable must be less than 10 feet
From +12 to R11A/D2A and common to common between Fig 11-18 and Fig. 11-19
Use #20 hook up wire and twist
From R47 and R52 to R49 and R53 respectively use #24 hook-up wire and twist
From TE4 #3 to M1 and M2 common (use #24 hook-up wire)
From T3B to D10B and D12B on Fig. 11-18

Wires Referenced Fig. 11-31

From TE4 #3 to R61 and R57 (use #24 hook-up wire)
From R56 and R60 to R57 and R61 respectively (use #24 hook-up wire and twist) on Fig. 11-17
From R57 and R61 arm to R1A and R2A respectively (use #24 hook-up wire and twist) on Fig. 11-19
From D1A to Q1A and Q2A (use #24 hook-up wire and twist) on Fig. 11-19

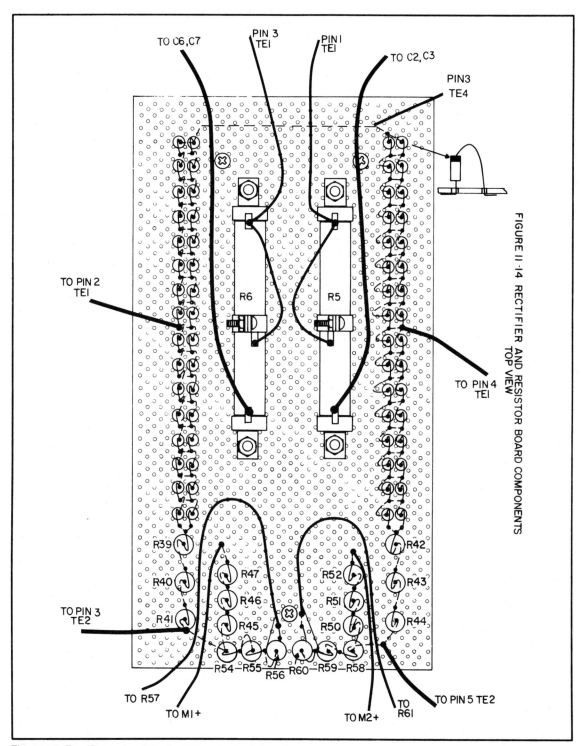

Fig. 11-17. Rectifier and resistor board components top view.

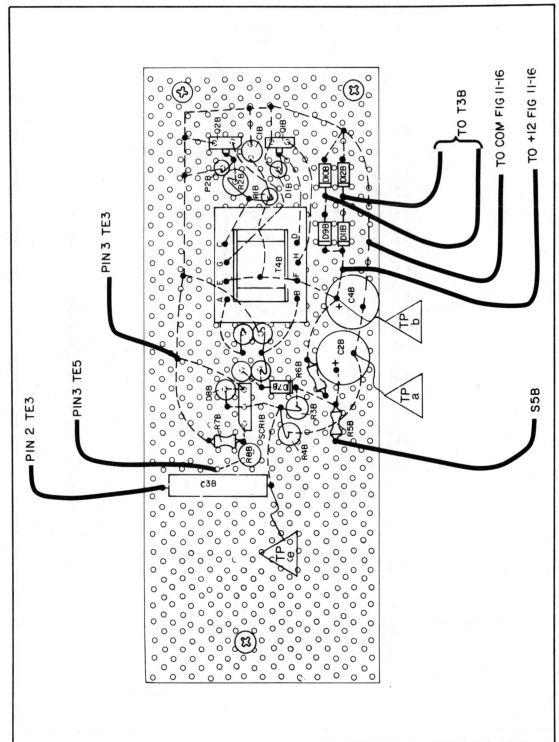

Fig. 11-18. Ignitor/12 Vdc PWR board components; top view.

152

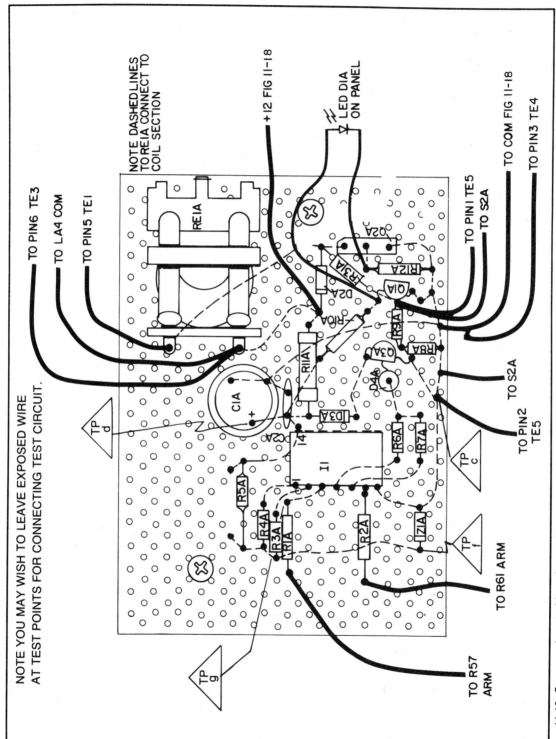

NOTE YOU MAY WISH TO LEAVE EXPOSED WIRE
AT TEST POINTS FOR CONNECTING TEST CIRCUIT.

NOTE DASHED LINES
TO REIA CONNECT TO
COIL SECTION

TO PIN6 TE3

TO LA4 COM

TO PIN5 TEI

+12 FIG II-18

LED DIA
ON PANEL

TO PIN I TE5

TO S2A

TO COM FIG II-18

TO PIN3 TE4

TO S2A

TO PIN2
TE5

TO R6I ARM

TO R57
ARM

Fig. 11-19. Sense and control board components; top view.

153

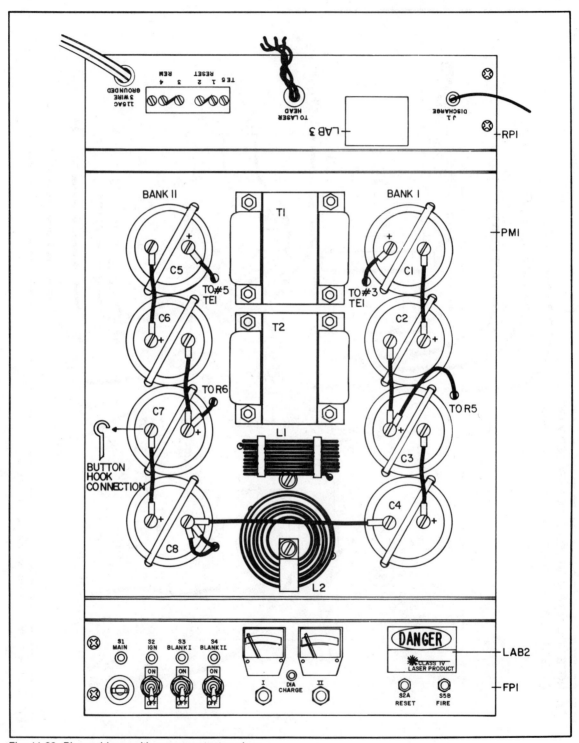

Fig. 11-20. Plywood base with components; top view.

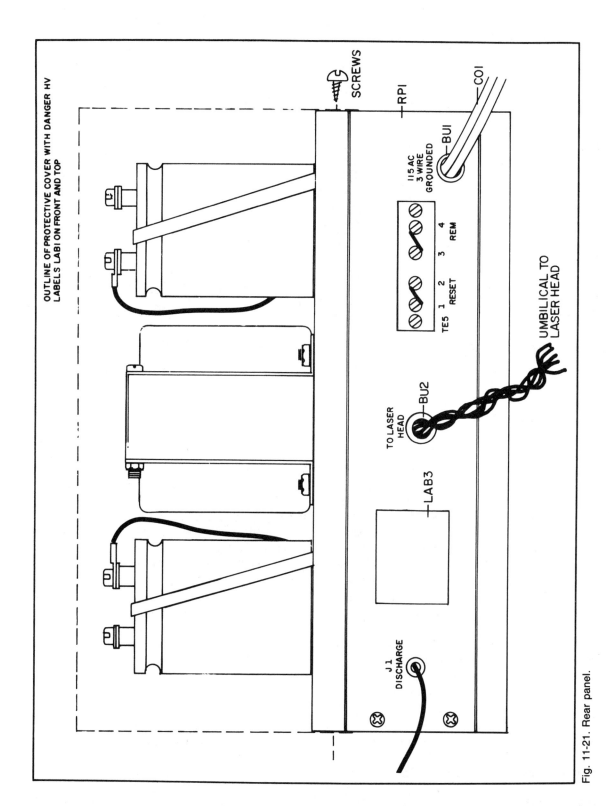

OUTLINE OF PROTECTIVE COVER WITH DANGER HV
LABELS LAB1 ON FRONT AND TOP

SCREWS

RPI

COI

BUI

115 AC
3 WIRE
GROUNDED

TE5 1 2 3 4
RESET REM

BU2

TO LASER
HEAD

UMBILICAL TO
LASER HEAD

LAB3

J 1
DISCHARGE

Fig. 11-21. Rear panel.

155

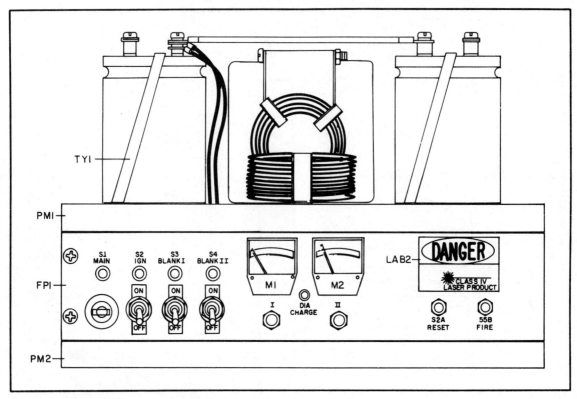

Fig. 11-22. Front panel.

duty nylon tye wraps or other nonconducting material. It is suggested to attach the front and rear panel sections as shown in Fig. 11-23 and layout the components as shown in Fig. 11-20 , and then drill the necessary holes.

7. Assemble all front and rear panel components along with capacitors and finally T1 and T2 transformers. Wire C2-C8 capacitors together as shown in Fig. 11-20. Use #14 buss wire for short jumps from plus to minus between C8 to C7, C7 to C6, C6 to C5, C4 to C3, C3 to C2, C2 to C1. Note that lugs are shown but may be replaced by a "button hook" connecting these solid wires. See Figs. 11-24 and 11-25.

8. Wind 45 turns of #12 magnet wire on a 2 inch form as shown in Fig. 11-20. Hold it together with nylon tye wraps. Allow 10-inch leads for interconnection. These are the inductors L1 and L2.

9. Proceed to wire in order as shown by Figs. 11-26 through 11-32 using the appropriate wire chart tables. Note that the sketches show the wire routes for clarity. It is suggested that you layout, measure, and neatly bundle individual wires with tye wraps and secure with nylon clamps etc. (A wire harness makes a neat looking layout.) See Figs. 11-33 through 11-40.

Discharge wiring should be direct and not included into the cable bundle. Assembly boards should not be fastened into position until all wiring is completed since access to their underside is required. Double-check connection points as errors at this point can be costly and time consuming.

10. Double-check all wiring, soldering, etc.

11. Assemble and attach safety discharge probe as shown in Fig. 11-25. Double-check the wiring of R66 and R67.

12. Attach *Danger High Voltage* labels to the following places:

    A. On T1 (Fig. 11-25).
    B. Top and front of protective cover (Fig. 11-24).
    C. On side of C4 and C8 for maximum visibility (Fig. 11-25).
    D. Underside as shown in Fig. 11-26(A).

At this point it is assumed that the power supply section of your ruby laser project is completed and ready for testing.

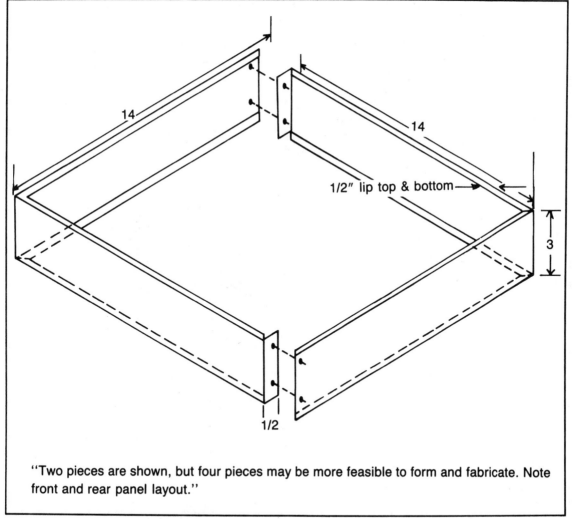

"Two pieces are shown, but four pieces may be more feasible to form and fabricate. Note front and rear panel layout."

Fig. 11-23. Rear and front panel fab suggestion.

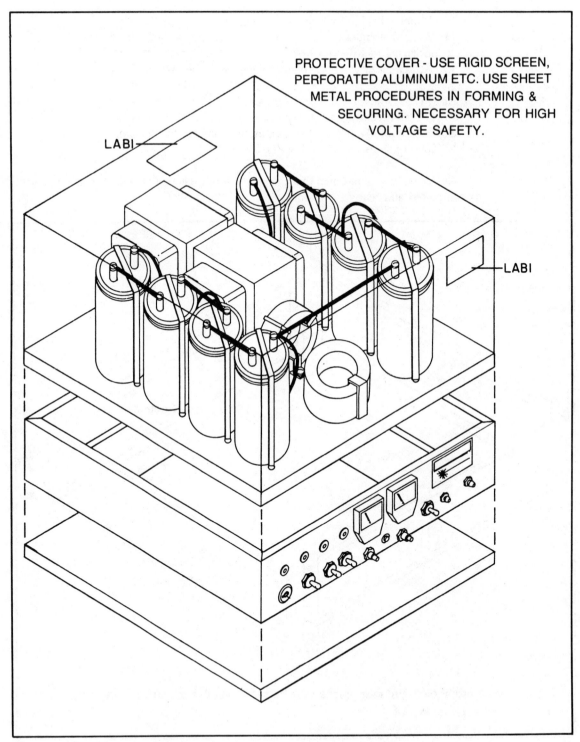

PROTECTIVE COVER - USE RIGID SCREEN, PERFORATED ALUMINUM ETC. USE SHEET METAL PROCEDURES IN FORMING & SECURING. NECESSARY FOR HIGH VOLTAGE SAFETY.

LABI

LABI

Fig. 11-24. Exploded view showing protective cover.

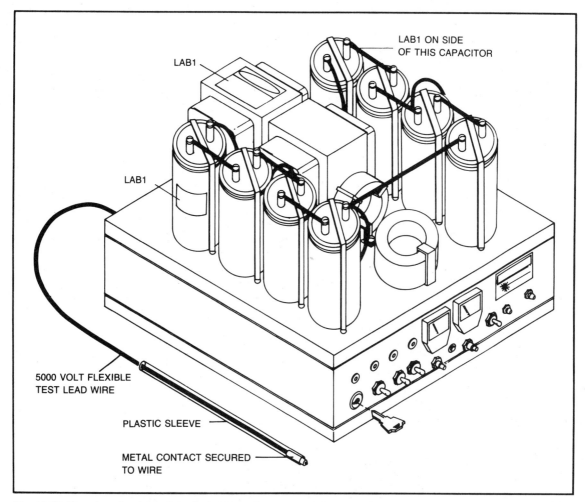

LAB1 ON SIDE
OF THIS CAPACITOR

LAB1

LAB1

5000 VOLT FLEXIBLE
TEST LEAD WIRE

PLASTIC SLEEVE

METAL CONTACT SECURED
TO WIRE

Fig. 11-25. Overview showing safety discharge probe.

It is unfortunate that to operate this laser system, dangerous and deadly high voltages are necessary. We cannot stress the importance of using extreme caution at this point as the voltage and energy levels if improperly contacted to the human body can cause electrocution or severe burns.

Several important guidelines that we demand for your personal safety are the following:

1. Work on a dry wooden floor. Never on a cement or cellar floor.

2. Always individually discharge each storage capacitor with a heavy insulated screwdriver even after using the safety probe even if the panel meters read zero volts. Your life may depend on this step.

3. Keep one hand in your pocket when working with the system energized. This helps to prevent electrical current from flowing through the main torso section of the body.

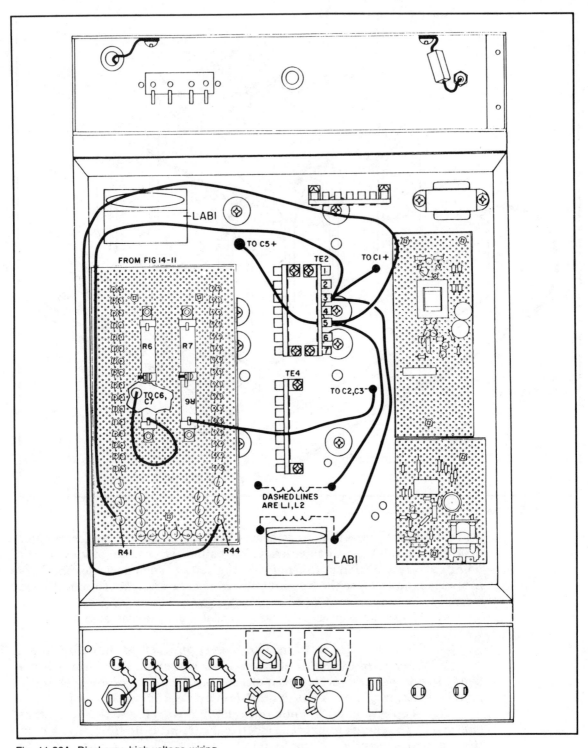

Fig. 11-26A. Discharge high voltage wiring.

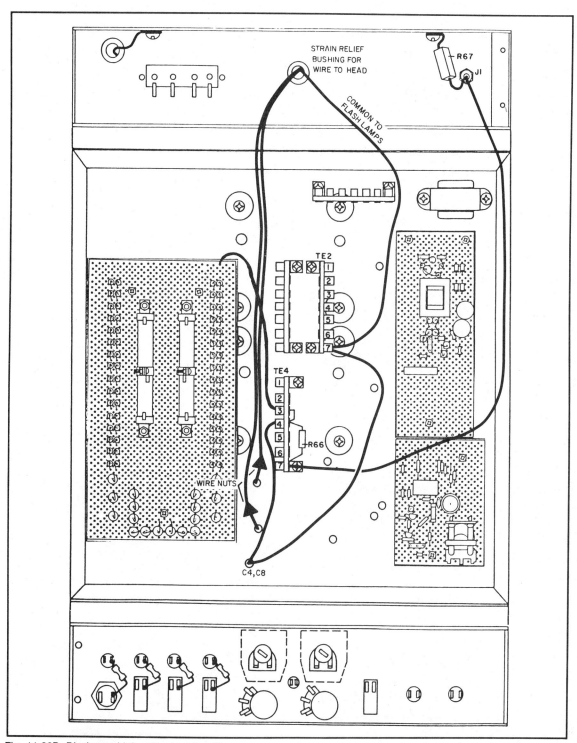

Fig. 11-26B. Discharge high voltage wiring #2.

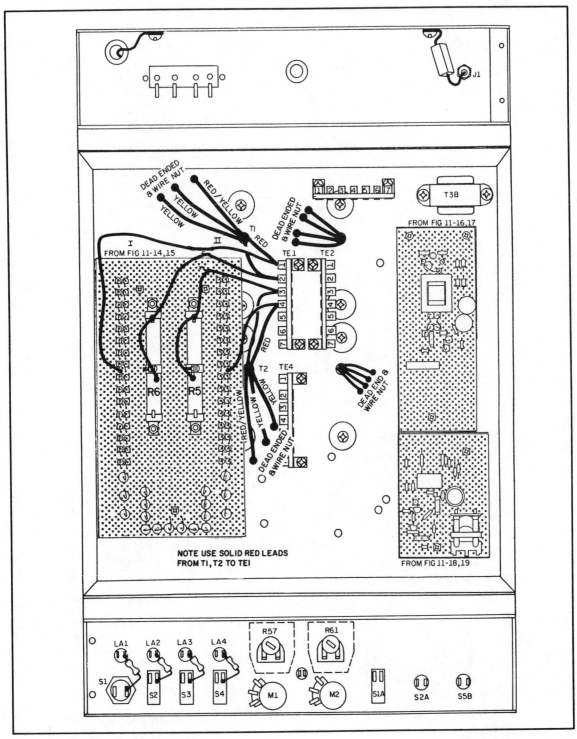

Fig. 11-27. Transformer secondary wiring.

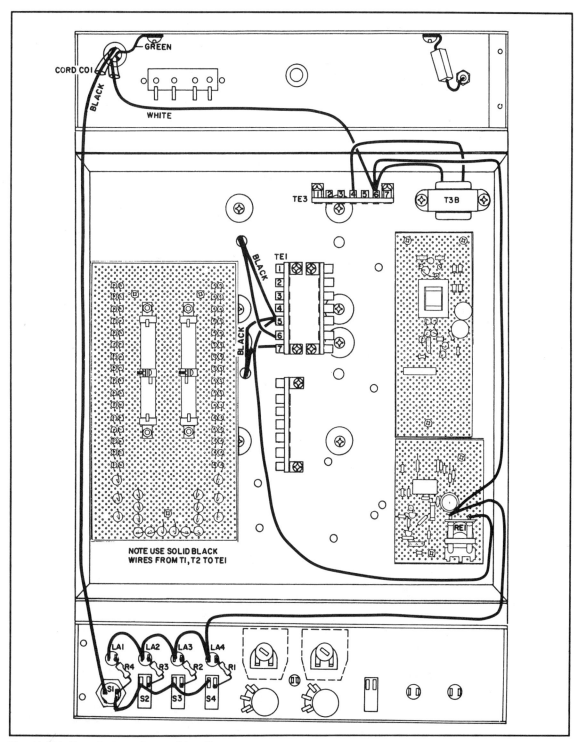

Fig. 11-28A. Transformer primary wiring #1.

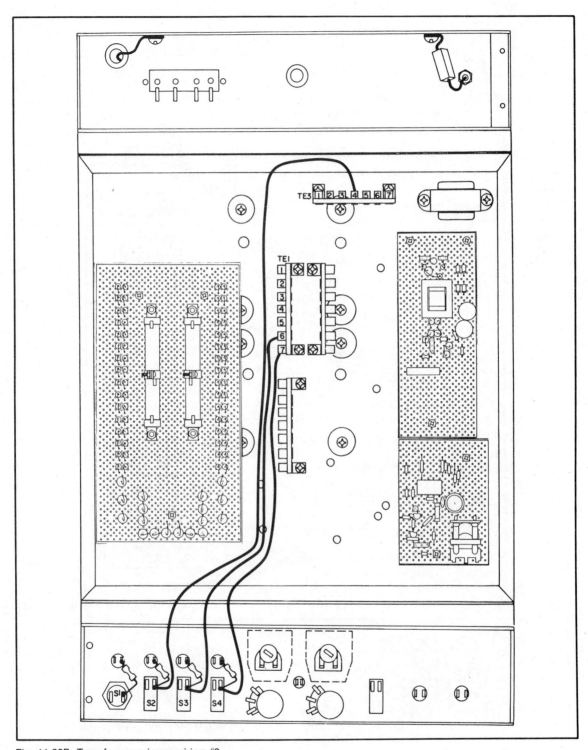

Fig. 11-28B. Transformer primary wiring #2.

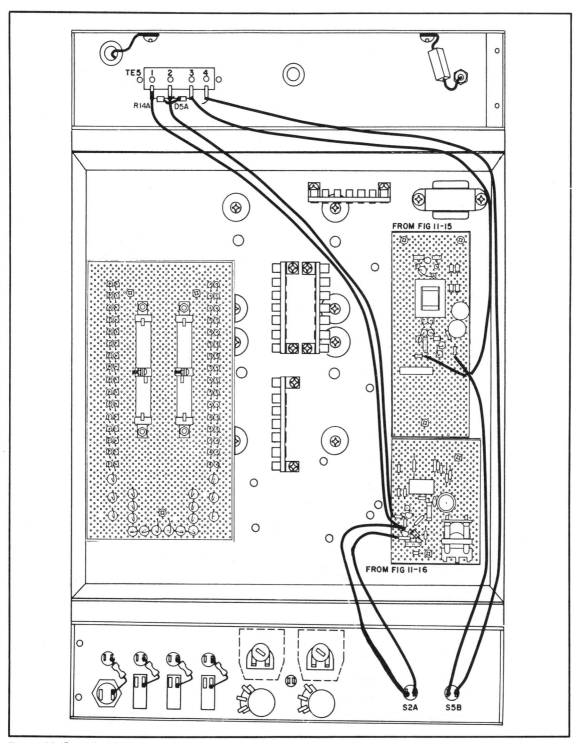

Fig. 11-29. Control wiring.

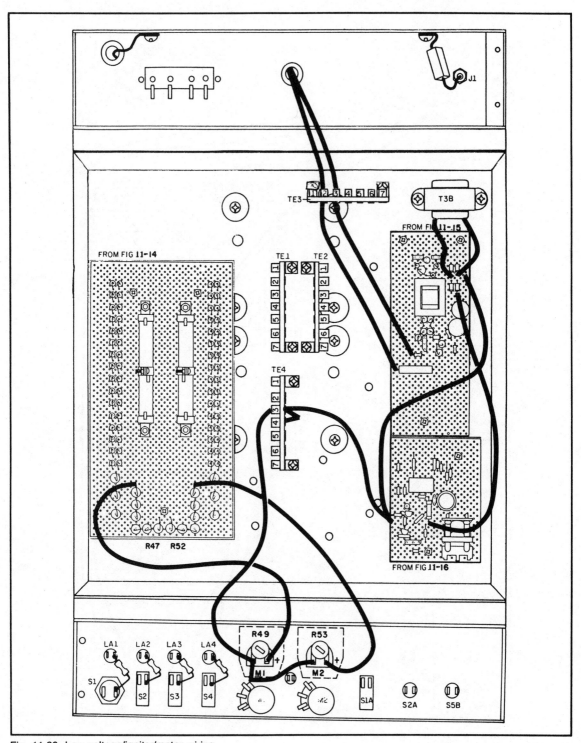

Fig. 11-30. Low voltage/ignitor/meter wiring.

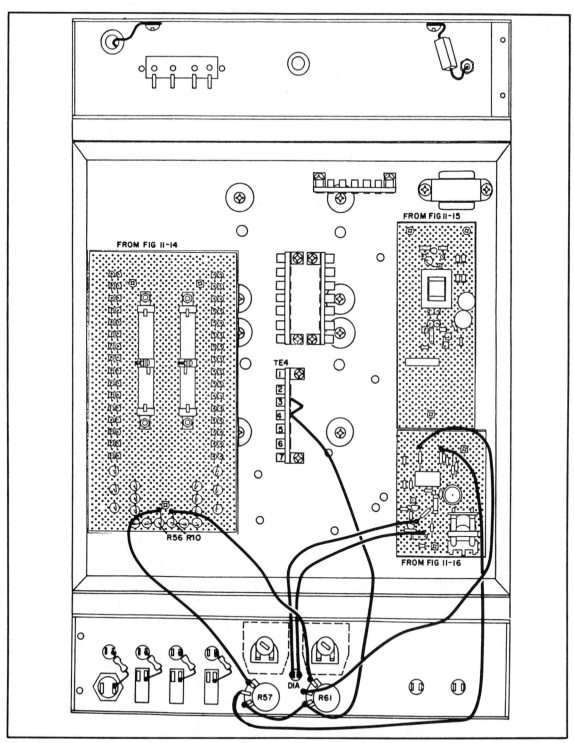

Fig. 11-31. Voltage sense wiring.

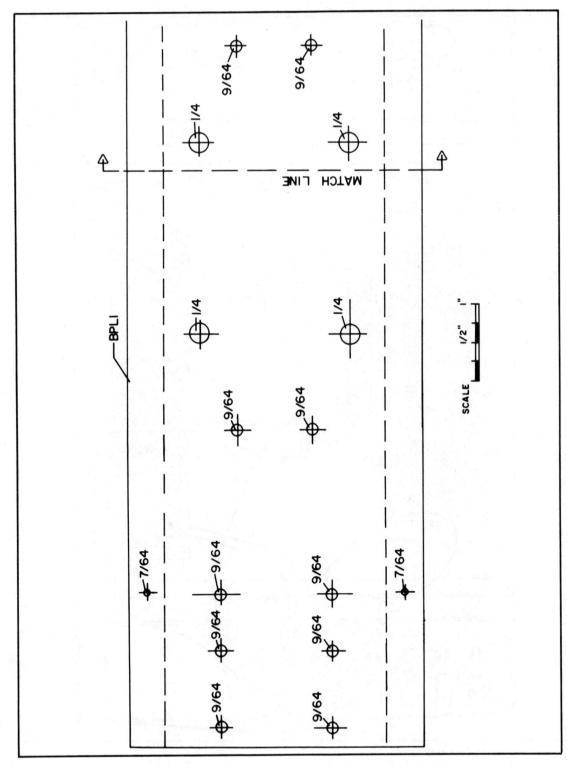

Fig. 11-32A. Rub 50 laser head bottom plate layout (left side).

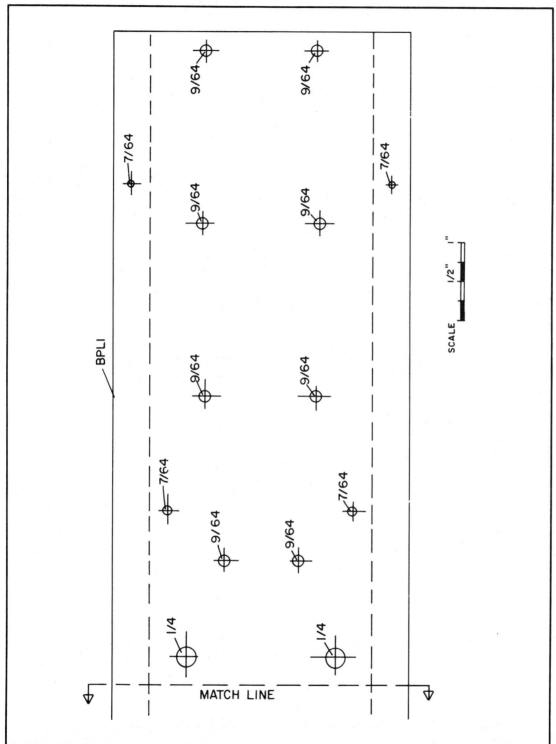

Fig. 11-32B. Rub 50 laser head bottom plate layout (right side).

169

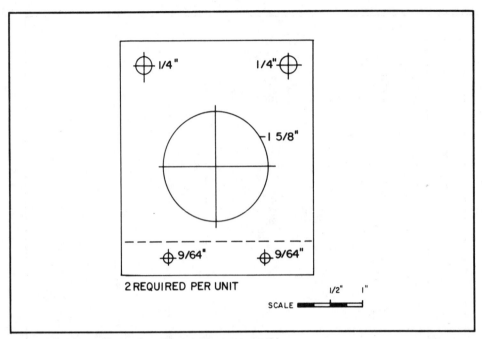

Fig. 11-33. EPL1-exit panels to house colimator assembly.

4. **Have a buddy present, keep alert, sober, and cautious.**
5. **Follow labelling instructions.**
6. **Have a complete understanding of the circuitry and all of the switching and related functions.**

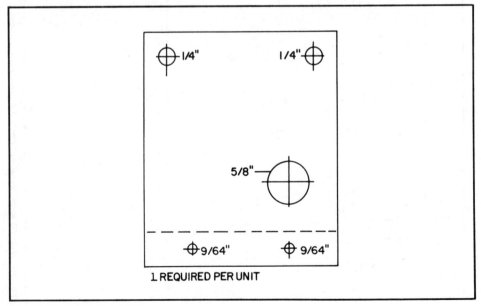

Fig. 11-34. END1-rear panel.

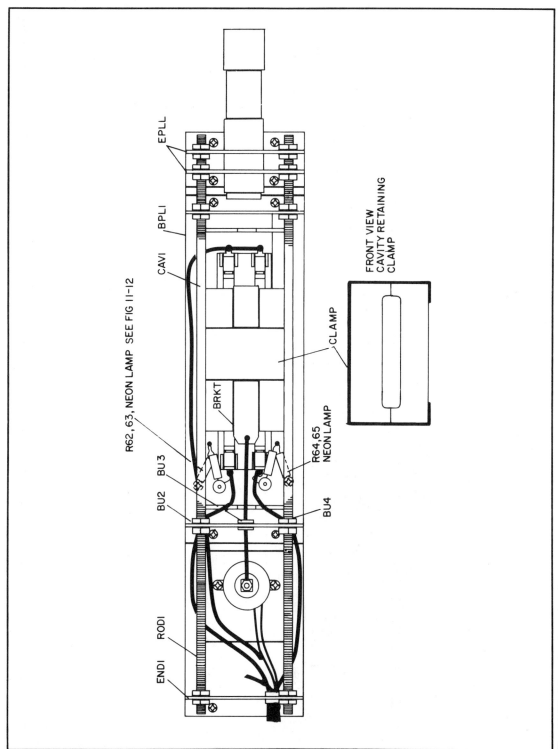

EPLL

BPLI

CAVI

R62,63,NEON LAMP SEE FIG 11-12

BRKT

BU3

BU2

BU4

R64,65
NEON LAMP

CLAMP

FRONT VIEW
CAVITY RETAINING
CLAMP

ENDI RODI

Fig. 11-35. A top view laser head.

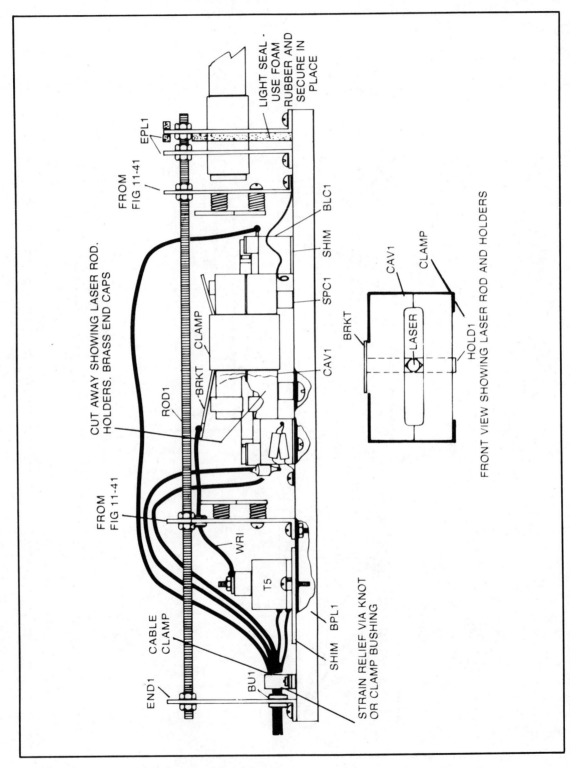

CUT AWAY SHOWING LASER ROD. HOLDERS. BRASS END CAPS

FROM FIG 11-41

FROM FIG 11-41

LIGHT SEAL - USE FOAM RUBBER AND SECURE IN PLACE

EPL1

ROD1

BRKT CLAMP

END1

CABLE CLAMP

WRI

T5

BU1

SHIM BPL1

STRAIN RELIEF VIA KNOT OR CLAMP BUSHING

CAV1

SPC1

SHIM BLC1

BRKT

CAV1

LASER

CLAMP

HOLD1

FRONT VIEW SHOWING LASER ROD AND HOLDERS

Fig. 11-36. Internal view.

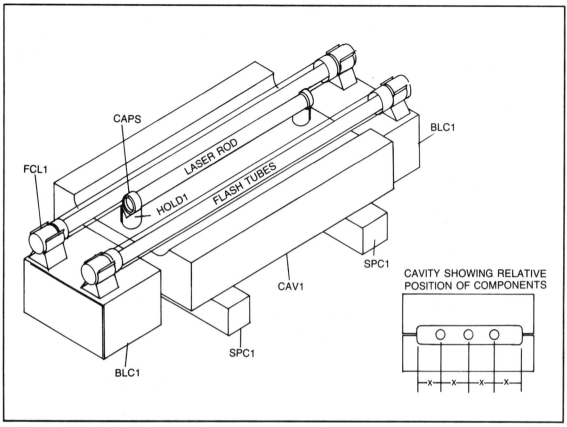

Fig. 11-37. Bottom half of cavity showing flash lamps and laser rod.

> **7. Have a good understanding of the type of electronic circuitry before even thinking to attempt this project.**

TEST PROCEDURES

1. Turn all switches off. Connect T5B primary to appropriate wires as shown in Fig. 11-10.

2. Securely connect a clip lead across both capacitor banks. This is a safety step to avoid premature charging of the capacitors.

3. Place a piece of insulative material between both contacts of RE1.

4. Obtain a 100-watt lamp (used for a ballast in case of gross short circuits) and insert across contacts of key switch S1 (make sure this switch is off).

5. Plug in line cord and note LA1 igniting. The 100-watt lamp should not show any glow.

6. Energize S2 and note LA2 igniting. Measure 12-Vdc across C4B and a slightly lower value across C1A.

7. Measure approximately 200 volts at junction of R3B and R4B to common.

8. Connect clip lead to pins 3 and 4 of TE5.

9. Energize S5B and note a high voltage pulse occurring between the output ter-

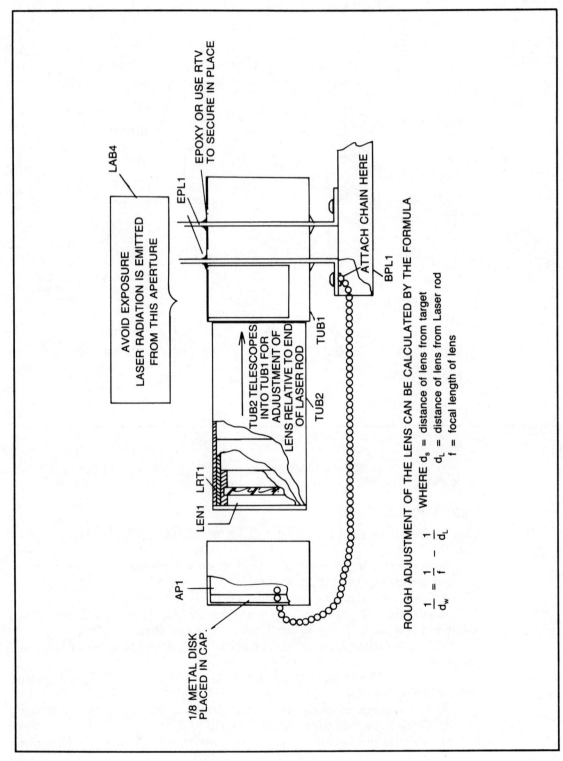

Fig. 11-38. Lens mounting showing aperture cap and retaining chain.

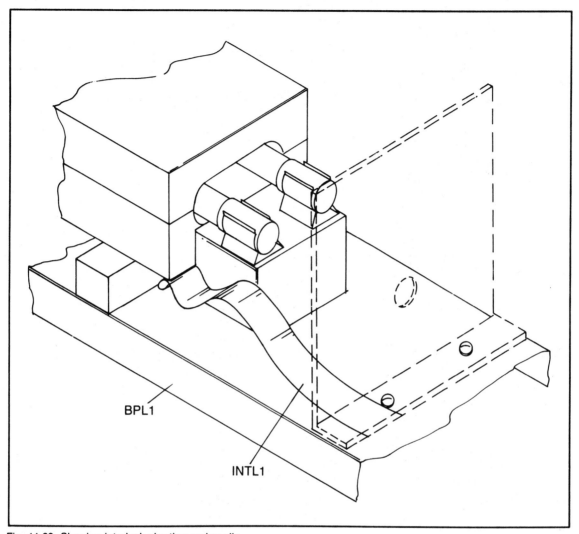

Fig. 11-39. Showing interlock shorting spring clip.

minals of T5B. Check for delay time necessary before a second pulse is obtained. Remove jumper across pins 3 and 4 of TE5 and note disabling of high voltage pulse when S5B is energized. Then check the delay, ignition, and remote control functions. Finger touch power tabs of Q1B and Q2B and note if they are only warm. Allow to remain in this mode for several minutes and check for any overheating of components.

10. Push S2A and note relay RE1A energizing. Connect jumpers between Z1A and R1A and note RE1A deenergizing.

11. Turn on S1A and press S2A—note RE1A holding. Turn of S1A then turn on.

12. Repeat by shorting pins 1 and 2 of TE5 noting RE1A holding. Note light-emitting diode D1A igniting indicating RE1A is energized.

13. Turn on S3 note LA3 igniting.

14. Turn on S4 note LA4 igniting. Note that these lamps indicate these circuits

are energized even though the actual transformers are not, due to the insulation material placed between RE1A contacts.

15. Turn S3 and S4 off. Remove material from between contacts of RE1A.

16. Momentarily turn S3 on and note the 100-watt lamp glowing to almost full brilliance. Repeat with switch S4. Return both S3 and S4 to off.

17. Remove lamp connected across S1 and note LA1 and LA2 extinguishing. Turn key switch (S1) on only momentarily and note these lamps igniting. Note that capacitor banks are still shorted and an overload may result.

18. Return all switches to "off" and remove all clip leads and power plug.

19. Set R5 and R6 for maximum resistance. These may be changed for faster charging, however, we recommend leaving them at maximum.

20. Carefully connect a 2000-volt meter across C1 to C4 in Bank I.

21. Connect ballast lamp as in previous step and turn "on" S3. Press S2A and note the voltmeter starting to indicate a charging action on the capacitors. Lamp should be glowing. Allow to charge to 500 volts and then turn "off" S3. Discharge all capacitors and repeat for Bank II energizing S4. Turn "off" all switches. *Completely discharge capacitors as previously outlined.* Note that the relay RE1A may not properly latch and require mechanical actuation using a pencil etc., for this step.

22. Remove ballast lamp and turn on key switch S1. Repeat previous step allowing individual banks to charge to 1500 volts. Note charge time being considerably faster with the ballast removed. It may be necessary to reset R57 and R61 to keep RE1A energized. The setting of these pots determines the amount of charge voltage before the comparator sends its turn-off pulse to RE1A terminating charging action. It may be necessary to adjust these to allow RE1A to remain energized until the 1500-volt levels are determined. When S1A is "on" RE1A stays deenergized after it reaches its preset value and remains "off" until the system is reset by actuating S2A. When S1A is off RE1A will alternately reenergize and deenergize to keep the capacitors charged up. S2A also provides the necessary reset required by the BRH should power be interrupted. Note D1A indicating when system is charging. Adjust trimpots on M1 and M2 to provide an indication corresponding to 1500 volts. This is usually the red band at the 2/3 point on the meter scales. **Use caution not to contact any high voltage points when making the adjustments to R49 and R53.** This completes the Power Supply Section.

BRIEF DESCRIPTION OF LASER HEAD

Your laser head is shown constructed using a 3 inch rod and flash lamps. Optional adjustable external mirror mounts are shown in Figs. 11-41 and 11-42 for those desiring a more flexible approach. This section is attached to the power supply via an umbilical cord that must not exceed ten feet in length for compliance with the BRH. The assembly is shown built on a metal channel base plate as shown in Fig. 11-32(A) and (B). Hole dimensions and a scale factor are shown, however, it is suggested to trial position all components before actual drilling of the channel. A protective cover also required by the BRH is shown in Fig. 11-40. It is fabricated from #24 galvanized sheet metal and secured via screws to the metal base plate. Figure 11-38 shows a simple lens system with aperture cover and retaining chain. Neon lamps and associated dropping resistors are connected across each flash lamp to provide a warning to the user

176

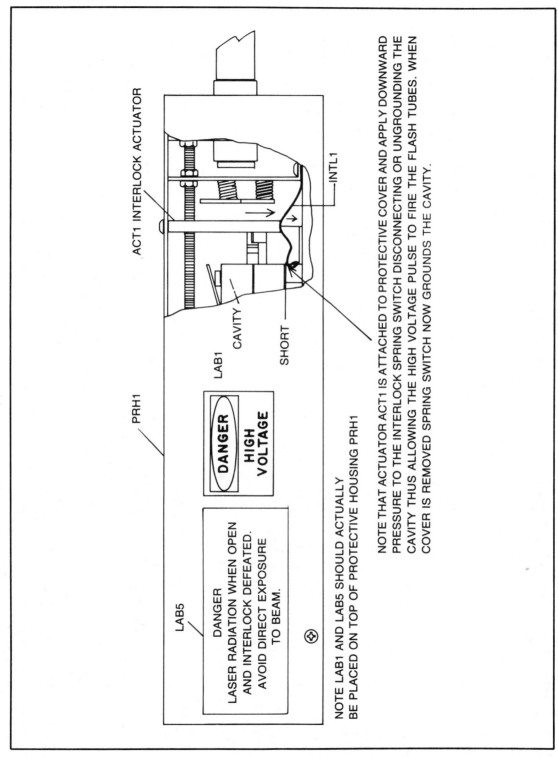

Fig. 11-40. Side view interlock switch.

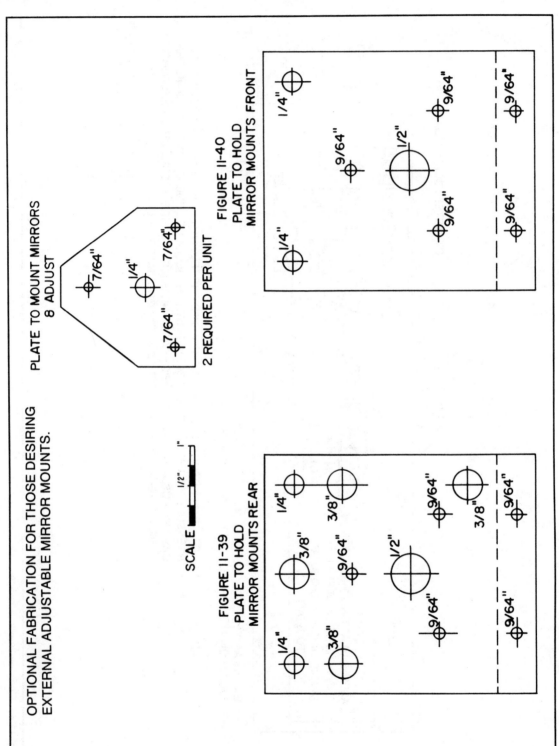

Fig. 11-41. Externally adjustable mirror mounts.

178

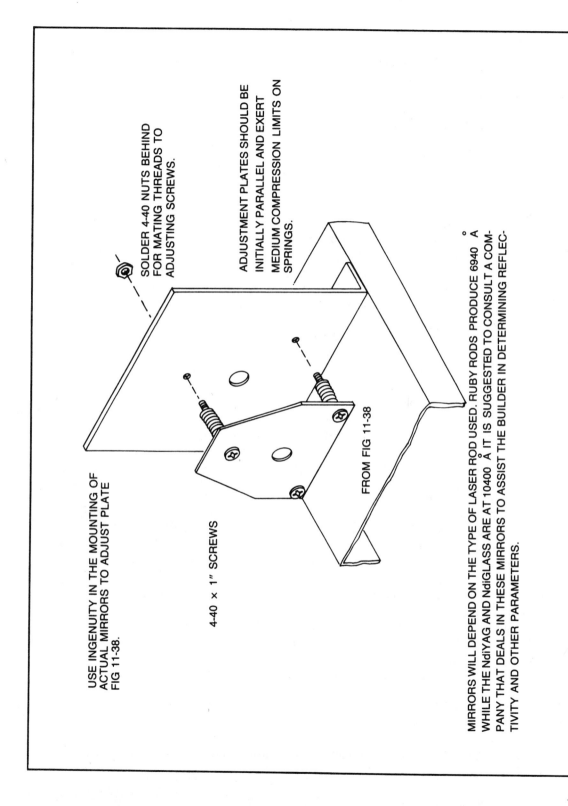

SOLDER 4-40 NUTS BEHIND FOR MATING THREADS TO ADJUSTING SCREWS.

ADJUSTMENT PLATES SHOULD BE INITIALLY PARALLEL AND EXERT MEDIUM COMPRESSION LIMITS ON SPRINGS.

USE INGENUITY IN THE MOUNTING OF ACTUAL MIRRORS TO ADJUST PLATE FIG 11-38.

4-40 × 1" SCREWS

FROM FIG 11-38

MIRRORS WILL DEPEND ON THE TYPE OF LASER ROD USED. RUBY RODS PRODUCE 6940 Å A WHILE THE NdiYAG AND NdiGLASS ARE AT 10400 Å IT IS SUGGESTED TO CONSULT A COMPANY THAT DEALS IN THESE MIRRORS TO ASSIST THE BUILDER IN DETERMINING REFLECTIVITY AND OTHER PARAMETERS.

Fig. 11-42. Mirror adjustment plate.

179

of the ever impending danger of the high voltage that remains on the capacitors even when power is turned off.

The cavity is isolated above ground and is energized by a HV pulse from T5 to capacitively trigger the flash lamps. A safety interlock is shown that defeats the trigger pulse when the protective cover is removed.

LASER HEAD CONSTRUCTION STEPS

1. Fabricate BPL1, bottom base plate from #22 to #24 galvanized stock. Use Fig. 11-32(A) and (B) as a starting guide but it is suggested to layout all other components before final drilling, etc.

2. Fabricate EPL1 (2) per unit, END1 as shown in Figs. 11-33, and 11-34. use #22 to #24 gauge galvanized stock.

3. Fabricate PRH1 from #24 gauge galvanized stock as shown in Fig. 11-40. This housing must be lightproof for compliance with the BRH. Note rubber abutment seals where cover is adjacent to RPL1 and END1 end Fig. 11-36.

4. Fabricate INTL1 switch from a piece of spring beryllium copper as shown in Figs. 11-36 and 11-40. The objective of this part is to ground the cavity to the metal base section, thus preventing the cavity from being triggered by the high voltage pulse. When the protective cover is now in position a piece of insulative material (ACT1) now provides actuation of the spring, disconnecting it from the cavity. Note that the separation with cover in place must be at least 3/8 inch. Shape and form to perform the above functions. This is a requirement by the BRH and requires the appropriate interlock labelling.

5. Fabricate the clamp to hold the cavity together as shown in Figs. 11-35 and 11-36 from a piece of #24 gauge metal. Make it to fit the particular cavity dimension used. Note that it also retains BRKT bracket that in turn spring loads laser rod holders (HOLD1) together. Note connection to HV pulse transformer T5 via wire WR1.

6. Fabricate BRKT bracket as shown in Figs. 11-35 and 11-36. Adjust shape to particular cavity used.

7. Fabricate two 1/4-20 rods as shown in Figs. 11-35 and 11-36. Note nuts and washers used. This approach when finally assembled is reasonably rigid.

8. Fabricate AP1 aperture cap per Fig. 11-38. Use a 1 1/2 inch cap with a 1/8 metal disc placed as shown. Attach chain via solder. Caution cap must have metal disc to prevent accidental exiting of laser beam since power level will easily penetrate cap by itself.

9. Fabricate TUB1 from a 2-inch × 1 5/8-inch OD black PVC or aluminum extrusion as shown in Fig. 11-38.

10. Fabricate TUB2 from a 6-inch × 1-1/2 inch OD black PVC or aluminum extrusion as shown in Fig. 11-38. Note that TUB2 must telescope into TUB1 with some friction or a securing method must be used.

11. Fabricate BLC1 blocks from 1 1/4-inch × 1-inch × 3/4-inch teflon as shown in Fig. 11-37. These serve as supports for the clips to the flash lamps.

12. Fabricate SPC1 cavity spacers from 3-inch × 3/8-inch × 3/8-inch teflon as shown in Fig. 11-37. These are used to electrically isolate the metal cavity above ground for triggering reasons.

13. Fabricate HOLD from a 1/4-inch round teflon extension. Note "V" notch to

secure laser rod. Shape and size to position rod in center of cavity as per Fig. 11-36. Note BRKT applies spring loading to keep slight pressure on laser rod.

14. Fabricate ACT1 interlock actuator. Size to function as described in section on INTL1 interlock switch. Use suitable dry insulative material. Secure to protective housing by screw as shown in Fig. 11-40. Note that interlocks system is relatively fail-safe, because a positive abutting action is required for trigger enable.

15. Cavity-2 part 3 inch double ellipse silvered as shown in Fig. 11-37. This part is available from Information Unlimited, P.O. Box 716, Amherst, NH 03031. Write or call (603) 673-4730 for pricing and availability.

16. Layout and assemble as shown in drawings. Adjust cavity so that flash lamps and laser rod are perfectly centered. Note that flash lamps are equal distance from laser rod and cavity sides. See Fig. 11-37. Note shim stock may be required for vertical alignment for centering. Take caution when securing SPC1 to base. Do not use screws that will cause break down between cavity and metal channel base.

17. Fabricate small rings caps (CAP) from brass tubing of equal inner diameter to laser rod outer diameter. Caps should be about 1/4-inch long and slit so they can be pushed onto the laser rod without excessive force. Allow 1/8-inch protrusion over end of rod. This protects the integral mirrors from the direct flash of the lamps. Note that the slit section of the cap should be facing either directly down or up and be positioned on the Hold rod holders.

18. Wire and solder R62, R63, and LA5 to FLH1. Also wire and solder R64, R65, LA6 to FLH2. Note schematic Fig. 11-12 and mechanical positions shown in Figs. 11-35, 11-36. These are safety devices indicating presence of a charge voltage on the capacitor bank.

19. Secure lens LEN1 to TUB2 via retaining rings as shown in Fig. 11-38 or use your own ingenuity. Secure assembly to EPL1 as shown.

20. Attach umbilical cable to proper points as shown in Fig. 11-36. Strain relive wires via clamp or special bushing. Note insulation piece shown under T5.

21. Insert flash lamp into respective clips noting that they have polarity. Do not assemble other half of cavity.

22. Verify and check all wiring for strain relieving and accuracy.

23. Carefully position laser rod and secure second half of cavity as shown in Figs. 11-35, 11-36 and 11-37. It is assumed that the rod is a ruby rod and contains the integral mirrors negating any optical alignment of external mirrors. Make sure lamps and rod are free of dirt and oil from fingers etc. Use a soft tissue soaked in acetone. Do not use pressure on mirrors or they will scratch.

24. Select a target object preferably something black such as a piece of plastic, etc. Position in front of laser rod secured via a piece of modeling clay. Do not use the lens system at this time.

25. **Put on your safety glasses. make sure all personnel have adequate eye protection and proceed to fire up the system to a voltage level of 1500 volts per bank. Push "fire" button and note a bright flash and a loud "thud" type sound.**

26. Examine target for a blemish or burn mark about the size of the laser rod. This is laser output and must be focused to do useful work.

27. Position lens system as shown focused on target and repeat. You should now

actually blast a small hole in the target. Position dimensions can be obtained by the formula shown in Fig. 11-38. Modeling clay can be used to secure the target.

28. Note that the cavity should be cleared after every other shot as dust and contaminates can impede performance. **Use care and double-check voltage points by touching with the safety discharge rod before using your fingers because a severe shock hazard always exists on the flash lamps.**

29. Optional forced air cooling may be considered and is left up to the ingenuity of the builder. Such sources of air as vacuum cleaner outputs are easy to use and will supply adequate air to increase the repetition rate several times.

30. The output obtained with a Nd-glass rod is several times that of the ruby and obviously produces a greater effect. These are available with or without integral mirrors and can also be purchased from Information Unlimited, P.O. Box 716, Amherst, NH 03031. Write or call (603) 6763-4730 for pricing and availability.

31. Mirrors for antireflective coated rods may be obtained from Mells Griot and are easily aligned using a helium-neon laser such as that described elsewhere in this book. Again ingenuity is required.

32. Applications for the device are limited as the repetition rate is quite slow due to heat dissipation, etc. An excellent demonstration when using a ruby rod is to obtain a red balloon inside of a clear balloon and note the puncture of the inner one without damage to the outer. This may require some experimenting to perfect as both balloons are easily destroyed.

Chapter 12

**INVISIBLE LASER RADIATION –
AVOID EYE OR SKIN EXPOSURE
TO DIRECT OR SCATTERED
RADIATION**

40 watt CO_2
10.6 microns

CLASS IV LASER PRODUCT

High-Powered Continuous IR CO$_2$ Laser (LC7)

Caution—Use of controls or adjustments or performance of procedures other than those specified may result in a radiation hazard.

T HIS PROJECT DESCRIBES A MODERATELY POWERED CO$_2$ LASER CAPABLE OF PRO-
ducing a beam of heat at a wavelength of 10.6 microns with an energy of 30 +
continuous watts. This power level is capable of burning, cutting, fabbing, and weld-
ing thin sheet metal, plastics, wood, paper or just about any material that will absorb
this wavelength.

The laser is an axial flow device consisting of approximately thirty inches of active discharge length with a surrounding cooling jacket of water (see Fig. 12-1). Mirrors are internal cavity mounted on a flexible plate joined to the system via metal bellows.
This method eliminates output windows and still provides a convenient means of adjustment.

The project consists of two parts: the laser head section and the power supply.
These are shown connected to the necessary support equipment. Construction is straight-
forward with glass blowing and other specialized construction techniques minimized.
The approach produces a stable reliable device with limited frills and extras.

THEORY OF OPERATION

A carbon dioxide laser is by far one of the simplest lasing devices to assemble and
operate. This asset is sometimes a disadvantage as this energy can be very dangerous
in the hands of the inexperienced, both to the builder and his surroundings.

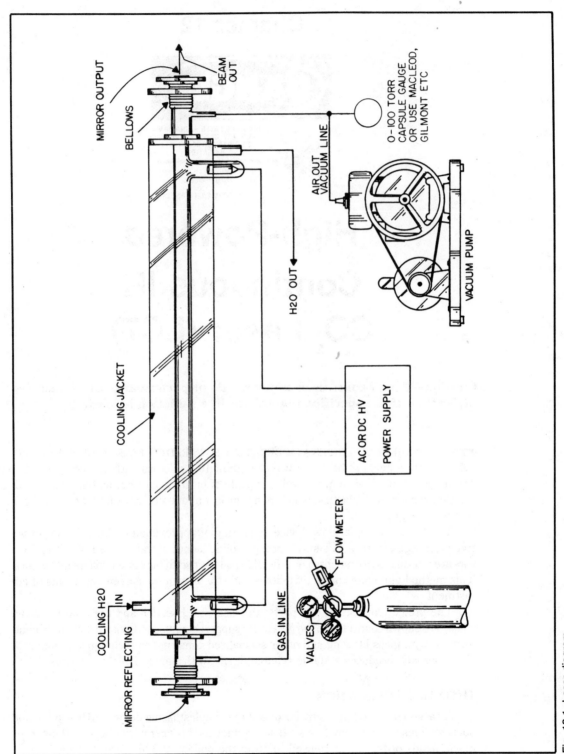

MIRROR OUTPUT

BELLOWS

BEAM OUT

AIR OUT VACUUM LINE

0-100 TORR CAPSULE GAUGE OR USE MACLEOD, GILMONT ETC

VACUUM PUMP

H2O OUT

COOLING JACKET

AC OR DC HV POWER SUPPLY

COOLING H2O IN

MIRROR REFLECTING

GAS IN LINE

FLOW METER

VALVES

Fig. 12-1. Laser diagram.

A CO_2 laser is a two-level, vibrational device emitting in the 10.6 micron infrared region. Lasing is accomplished by electrically exciting nitrogen gas N_2 to an energy level close to that of the CO_2 molecule (see Fig. 12-2). The main dissipation of this energy is in the resonant transfer to the CO_2 molecules causing them to change from

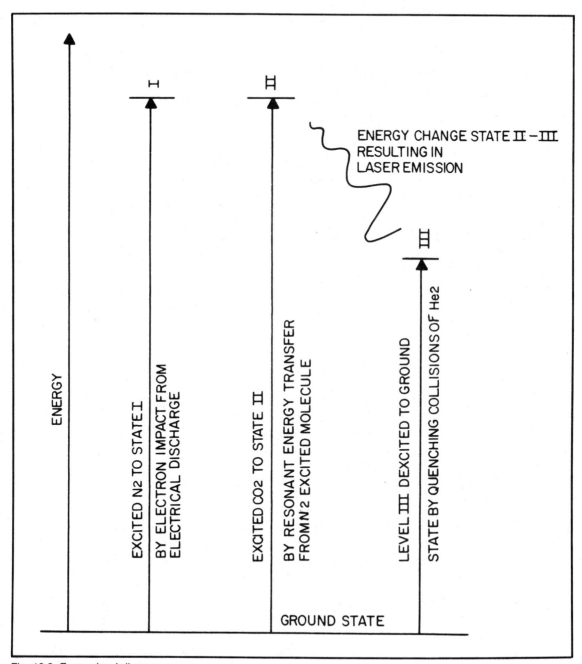

Fig. 12-2. Energy level diagram.

ground state to excited state 2. You will note that the lower laser level 3 is not prone to this transfer of energy, hence a population inversion now exists between levels 2 and 3. The laser transition commences as a result of the stimulated energy transition from level 2 to 3. A means now of returning level 3 energy to ground level is accomplished by quenching with a third gas (helium). Obviously, if level 3 was allowed to build up, the population inversion would soon be reduced between 2 and 3, hence terminating laser action.

It should be noted that the CO_2 gas molecule has many modes of vibration that contribute to the laser frequency. The main mode of vibration is around 10.6 microns and the system is often categorized in respect to this wavelength. Mother Nature's atmosphere offers a natural window for this wavelength, consequently a CO_2 laser is excellent in respect to applications requiring propagation over long distances. *A CO_2 laser of sufficient power is an excellent candidate for use as a direct energy beam weapon.*

The laser described in this project is referred to as a "low flow axial device." This is the simplest approach and is nothing more than an airtight glass discharge tube with gas being injected in one end and exiting the other via a vacuum pump. Electrodes now placed into the flowing gas excite the nitrogen molecules commencing laser action.

CONSTRUCTION SUGGESTIONS

Note that we have used standard readily available parts and pieces wherever possible while still providing reasonably reliable operation. It is suggested that the builder closely follow the plans. Specialized parts are available from Information Unlimited, as many hobbyists may not have the facilities for fabricating and machining.

Your CO_2 laser will be described in two sections: I. Laser Head, and II. Power Supply and Support Equipment.

The laser head section is where the action takes place. It is where the beam exits, and the lasing phenomenon occurs. Construction is started by fabricating the glass discharge tube with the builder having three choices regarding this step.

A. You may attempt to master glass blowing and build it yourself. A person interested in constructing lasers or plasma devices should seriously consider this craft along with acquiring the associated equipment necessary.

B. You may purchase the necessary glass pieces and contract an experienced glass blower such as those that do work in neon advertising signs.

C. You may purchase the ready made discharge tube from Information Unlimited as described in the parts list of these plans.

If you should attempt the tube yourself we offer the following basic glass blowing hints. It is also suggested that you acquire a manual or preferably a glass blowing kit available through VWR Scientific. This kit contains all the equipment and instructions necessary to do this project and, of course, much much more.

DISCHARGE TUBE GLASS BLOWING CONSTRUCTION STEPS

1. Clean the glass in hot water with a water softener. Hands must be kept clean as finger prints will easily "burn in" and dirty glass creates problems.

2. Refer to Fig. 12-3(A), cutting glass involves making a deep scratch using considerable pressure with a file where the break is to occur. Make the scratch with one

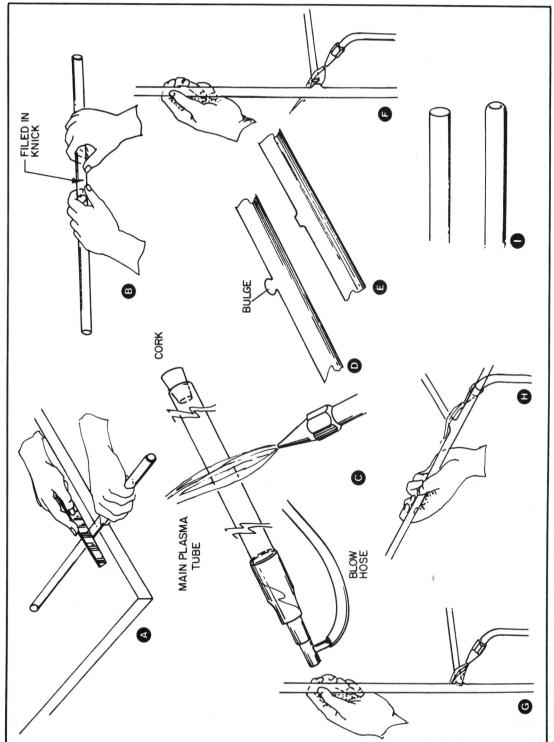

Fig. 12-3. How to cut and join glass tubing. See text for details.

187

pass. Wet the scratch and break tube as shown in Fig. 12-3(B). This method is usually good for diameters up to 1", therefore it suffices for these plans.

3. Learning to make a "T" seal. This is necessary for the joining of the premade glass electrodes to the main plasma tube. These seals are made by heating the desired location on the main tube using a small sharp flame positioned so that the top edge of the tube is just beneath the flame. After heating, gently blow a bulge that is slightly smaller in diameter than the tube that is to be sealed in. See Fig. 12-3(C) and (D). Reheat the top of this bulge and blow it away as shown in Fig. 12-3(E). Now hold this tube in the left hand between thumb and just three fingers so that the hole is to the left of the flame. Take the electrode tubing in the right hand and heat it uniformly until the end is soft and just before joining the two, heat the other opening uniformly. See Fig. 12-3(F).

Lift out of the fire and press the two together quickly, pulling slightly as soon as complete contact has been made. Blow slightly to expand heated area. If both edges are soft enough they will flow together upon contact and the small amount of blowing and pulling will give you a nice smooth seal as in Fig. 12-3(G). If, however, you find your seal heavy and uneven in wall thickness, heat one side at a time and blow gently only after the glass is sufficiently soft enough to work out the uneven area. See Fig. 12-3(H).

4. Fire polish the ends. This is simply accomplished by rotating the end of the cut tube in the flame until it is smoothed out by the surface tension. See Fig. 12-3(I).

This is a very, very brief section on glass handling and again it is strongly suggested that the laser builder seriously consider learning this useful art because many lasers and plasma devices involve glass handling in one way or another. The builder may wish to attempt the above steps several times until he feels proficient. Glass is not expensive and practice is advised.

LASER HEAD CONSTRUCTION STEPS

STEP 1. Construct or acquire the above Plasma Discharge Tube as shown in Fig. 12-4. Also, see Table 12-1.

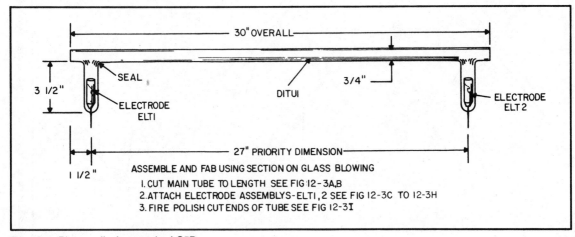

Fig. 12-4. Plasma discharge tube LC5D.

Table 12-1. Laser Head Parts List.

**MRMT1-Mirror Mount Sub Assembly Fig. 12-16.
  MMT1 (2) Fabricate per Fig. 12-9.
  MTGW1 (2) Fabricate per Fig. 12-12.
  BRWAS1 (2) Use as is.
    Above items fabbed from 2 1/4″ × 1/8″ brass washer with 15/16″ hole.

  ADWAS1 (2) Fabricate per Fig. 12-10.
  LAWAS1 (2) Fabricate per Fig. 12-11.
    Above items fabbed from 3 1/2″ × .155″ steel washer with 1 5/8″ hole.

ABP1	(2)	Fabricate per Fig. 12-7.
BELW1	(2)	Plated bellows cut in half.
O RING	(2)	1 7/8″ AS-032.
CPLS1	(2)	Fabricate per Fig. 12-13.
HNPS1	(2)	Fabricate per Fig. 12-14 or use 1/8″ × 1/4″ brass hose barbs.
ASW1	(6)	6-40 × 2″ adjusting screws.

**DITU1-Discharge tube assembly Fig. 12-4.

ELT1	(2)	200 mA sealed-in-pyrex electrodes.
SLV1	(2)	Fabricate per Fig. 12-8.

**COOL1-Cooling jacket assembly.

TROD	(4)	Threaded rods and nuts 1/4″ 20 × 36″.
CJ1	(1)	Fabricate per Fig. 12-5.
EB1	(2)	Fabricate per Fig. 12-6.
FM1	(4)	Fabricate per Fig. 12-15.
PC1	(1)	Protective housing Fig. 12-24. Use sheet metal as shown for both cover, bottom, and end pieces.
END1		End covers fab Fig. 12-24.
APC1		Aperture cap and coupling-use 1″ × 1″ copper tube and mating cap.
APR1		Aperture cap retaining chain.
BU1		HV lead retaining bushings.

**LABEL SET

LAB1,2,3	HV labels.
LAB4	Certification label.
LAB5	Class IV Type DD.
LAB6	Output Aperture Type E.
LAB7	Interlock defeat Type K without interlocks or use Type N with interlock.

**MIRRI	Mirror set consisting of resonator and output.
**MIRO1	Output mirror ZnSe 80% plano/plano 25 mm.
**MIRRI	Resonator mirror Si 99.5% plano/conc 25 mm × 10 meters.
LENS1	Focusing lens GaAs meniscus 25 × 125 mm.
ABP1	Modified from Fig. 12-7 to mount above lens.

STEP 2. Fabricate the cooling jacket (CJ1) from a 30″ piece of 3″ schedule 40 PVC as shown in Fig. 12-5. You will note the tendency for the slot to compress. This can be minimized by placing a 7/8″ block in the middle of the slot. This part is available all fabbed from Information Unlimited. Refer to the cooling jacket (Fig. 12-5). Note that the ends must be perfectly flush.

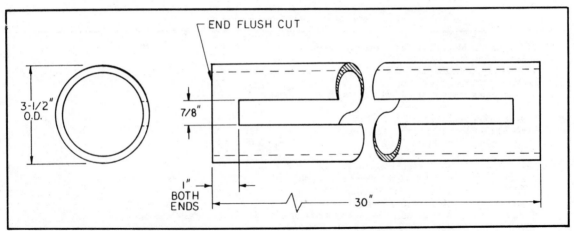

Fig. 12-5. Cooling jacket (CJ1).

STEP 3. Fabricate two end blocks (EB1) from preferably 1/2″ to 5/8″ PVC sheet as shown in Fig. 12-6. Note the mating groove for the cooling jacket. Roughen up groove with steel wool, etc.

STEP 4. Fabricate two abutting plates (ABP1) from 1/2″ to 5/8″ thick PVC sheet as shown in Fig. 12-7.

STEP 5. Fabricate two sleeves (SLV1) from 1″ OD × 3/4″ ID PVC schedule 40

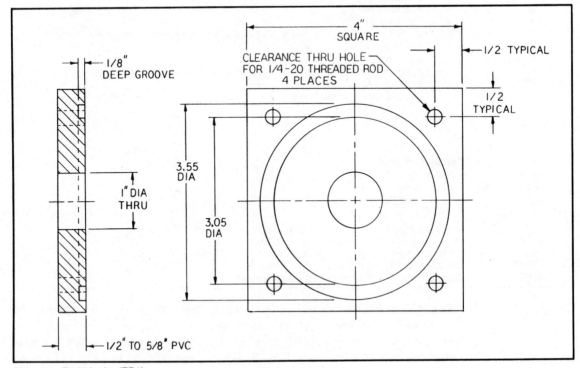

Fig. 12-6. End blocks (EB1).

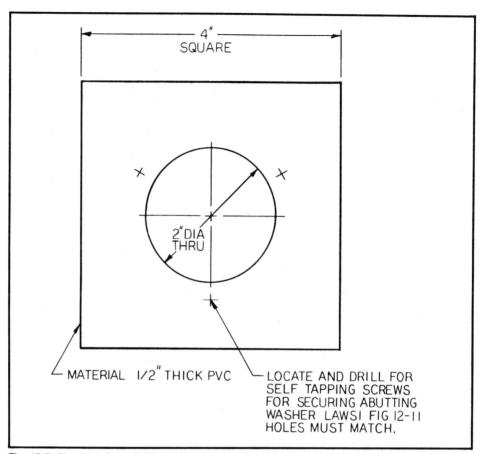

Fig. 12-7. Abutting plates (ABP1).

tubing as shown in Fig. 12-8. Note that the pieces must fit over the ends of the discharge tube and may need honing out for proper fit. Both inner and outer surfaces must be roughened up for proper bonding to the RTV (room temp vulcanize) sealant. Use steel wool or sandpaper.

STEP 6. Fabricate two mirror mounts (MMT1) from existing 2 1/2″ OD brass washers as shown in Fig. 12-9.

STEP 7. Fabricate two adjustment washers (ADWAS1) from existing 3 1/2″ steel washers as shown in Fig. 12-10. Note proper groove for "O" ring seat.

STEP 8. Fabricate two abutting washers (LAWS1) from existing 3 1/2″ steel washers as shown in Fig. 12-11.

STEP 9. Fabricate two mounting washers (MTGW1) from existing 2 1/2″ brass washers as shown in Fig. 12-12. *Use of existing washers for Figs. 12-9 through 12-12 eliminates most of the expensive machining for these components.*

STEP 10. Fabricate two coupling sections (CPLS1) from a 1 3/8″ OD copper tubing as shown in Fig. 12-13.

STEP 11. Fabricate two hose nipple sections (HNPS1) from 3/8″ copper tubing as shown in Fig. 12-14. Note that 1/8 × 1/4 brass hose barbs are recommended as these

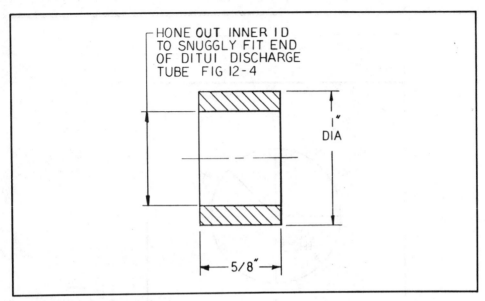

Fig. 12-8. Sleeve (SLV1).

more easily adapt to the vacuum hoses. Pieces must fit tightly into CPLS1 coupling section. Secure via soldering or epoxy.

STEP 12. Fabricate four frame members (FM1) from 46″ lengths of 1′ × 1′ square cornered aluminum angle as shown in Fig. 12-15(A).

STEP 13. Assemble mirror mount holders as shown in Fig. 12-15(B).

Assembly for this section will require a propane torch, flux and solder. Parts must

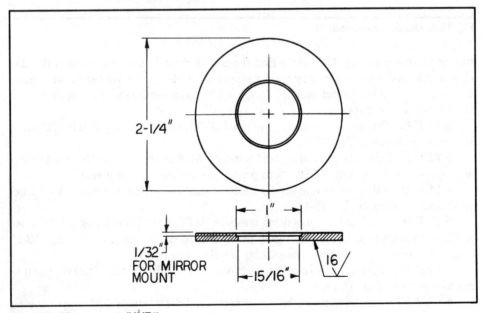

Fig. 12-9. Mirror mount (MMT1).

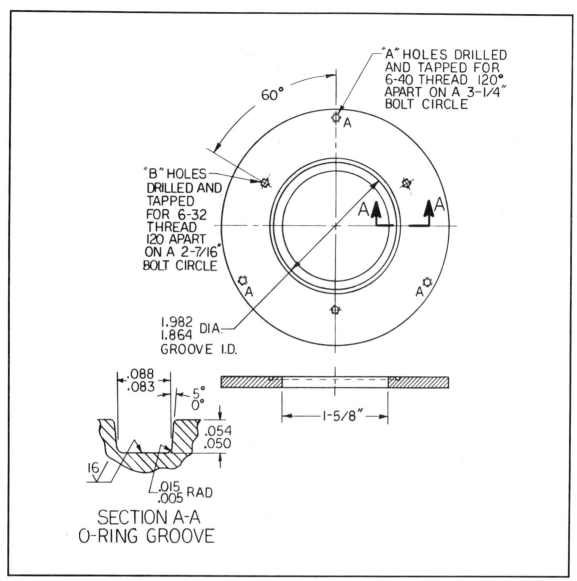

Fig. 12-10. Adjustment washer (ADWAS1).

be completely clean for proper bonding of the solder. Acid or rosin core can be used however parts must be thoroughly washed after bonding. Two complete assemblies are required.

A. Solder CPLS1 coupling section as shown in Fig. 12-13 to MTGW1 mounting washer as shown in Fig. 12-12. Make sure parts are well centered. Solder in copper tube HNPS1 shown in Fig. 12-14 while assembly is still hot. Note Fig. 12-16(D).

B. Solder BRWAS1 untouched brass washer to ADWAS1 adjustment washer as shown in Fig. 12-10. Keep centered. Note Fig. 12-16(A).

C. Obtain the bellows and cut in half with a sharp saw. File edges smooth and

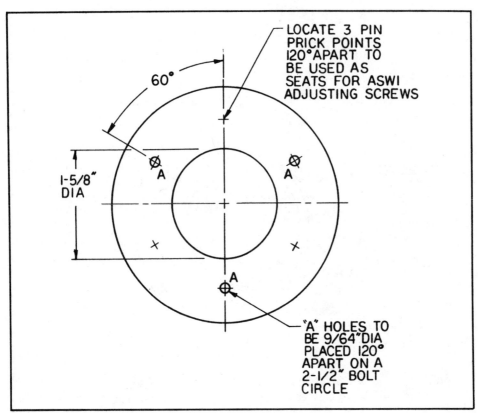

Fig. 12-11. Abutting washer (LAWS1).

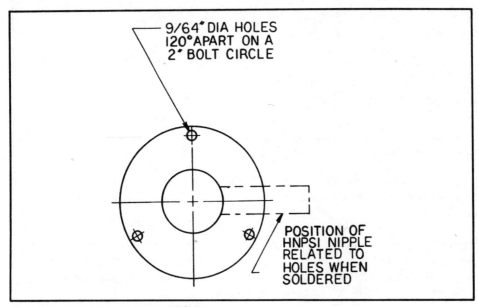

Fig. 12-12. Mounting washer (2) (MTGW1).

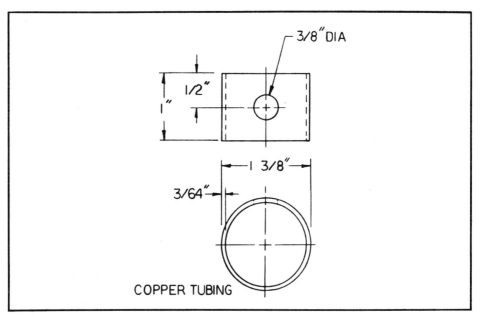

Fig. 12-13. Coupling sections (CPLS1).

clean. Cut bellows will have four to eight sections each if initial part is purchased from Information Unlimited.

D. Solder bellows to above BRWAS1 and ADWAS1 washer assembly in Step B. Note Fig. 12-16(A).

E. You now have the two halves of each assembly. Place the half with the bellows face down on the ADWAS1 washer. Slip the ABP1 abutment plate shown in Fig. 12-7 over the bellows. Solder section assembled in Step A to free end of bellows. Note not to burn the ABP1 PVC piece. Use a piece of metal for a heat shield. Note Fig. 12-16(C).

STEP 14. Insert discharge tube DIST1 into cooling jacket CJ1. Note that slot must be kept from compressing onto the glass and breaking discharge tube. A 7/8″ thick wooden block should be used as a spacer to keep the slot fully open during and

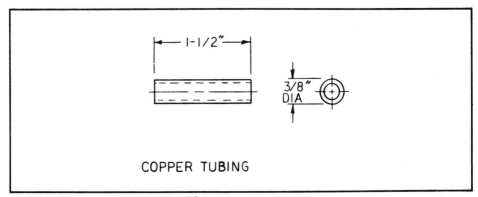

Fig. 12-14. Hose nipple sections (HNPS1).

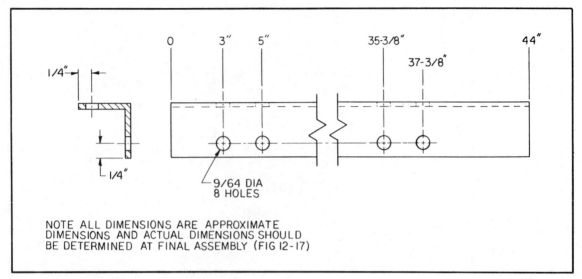

Fig. 12-15A. Frame members (4) (FM1 A,B,C,D).

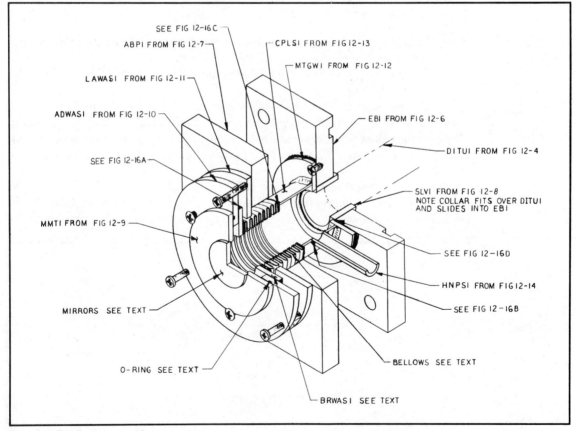

Fig. 12-15B. Mirror mount blow up.

after inserting tube. Place sleeves SLV1 on end of discharge tube and seal with RTV as shown. Make sure that the overall length with the sleeves will slightly recess into the PVC end block EB1 when they are secured to the cooling jacket as shown in Fig. 12-16.

Assemble EB1 end block to the above assembly using four long threaded rods. Run a bead of RTV in the circular slot for sealing the cooling jacket ends. Note that the sleeved ends of the discharge tube fit securely and evenly into the center holes of the end blocks and are slightly recessed for sealing with RTV. Attach FM1 frame members via screws as shown and properly position to align end blocks.

Secure assembly by evenly tightening the nuts on the threaded rods. Carefully "finger in" some RTV around the sleeves and the end block. Note Fig. 12-16(D). This must be perfectly sealed against any gas or liquid leakage. It is suggested to roughen up all PVC surfaces that are used for sealing with steel wool. Allow RTV to set and fill the cooling jacket with water to check for any leaks.

Attach mirror mount holders to end block via three screws. Roughen up surface of end block with steel wool and seal with RTV under MTGW1 mounting washer. Position ABP1 as shown and secure to frame member via screws. Note distance between the opposite faces of the adjustment washer and the abutting washers as this determines the length of the adjustment screws (ADSC1). Set these screws so that the adjustment washers ADWASI are nearly perpendicular to the bore of the tube with a slight loading on the spring action of the bellows. (Usually three complete turns after contact to the abutting plate.) See Fig. 12-17.

POWER SUPPLY CONSTRUCTION STEPS

Your laser head requires high voltages at relatively low current for operation. Raw ac is used in this case as it eliminates the use of rectifiers and helps reduce electrode heating especially at the positive end. While the beam may contain or be modulated by the line frequency, little effect as to the power output is noted. See Fig. 12-18 and Table 12-2.

POWER SUPPLY CIRCUIT THEORY

Transformer (T2) has its primary winding connected across variable voltage transformer (VA1). See Fig. 12-19. This approach allows continuous control of power to the system. The secondary of T2 is fed directly to the laser housing through the interlock scheme as shown. The interconnecting umbilical wires between the laser head and power supply are not to exceed three meters in length for compliance with this design. Secondary voltage is approximately seven to eight thousand volts once the plasma in the tube is sustained. Since the discharge through a gas is negative, that is, the dynamic resistance decreases with increasing current, a means of current control is required to prevent runaway. T2 provides internal current limiting, eliminating the use of a high-wattage resistor for ballast.

The primary circuit of T2 is energized via a contractor (K2) whose control coil is energized by relay (K1). A second set of contacts on K1 is used for a holding circuit keeping K1 energized until a power failure or the "stop switch" (S5) is actuated. A remote control terminal strip allows optional wiring of both the start/reset switch (S4) and stop switch S5.

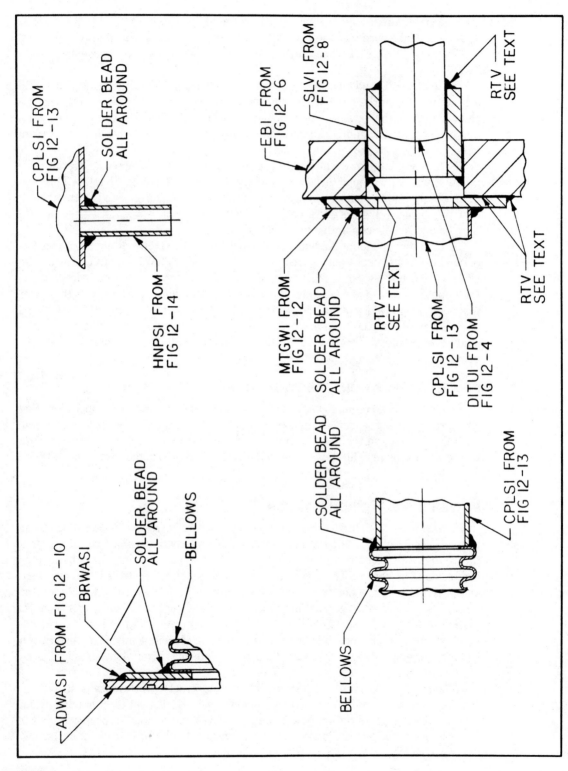

Fig. 12-16. Final assembly instructions.

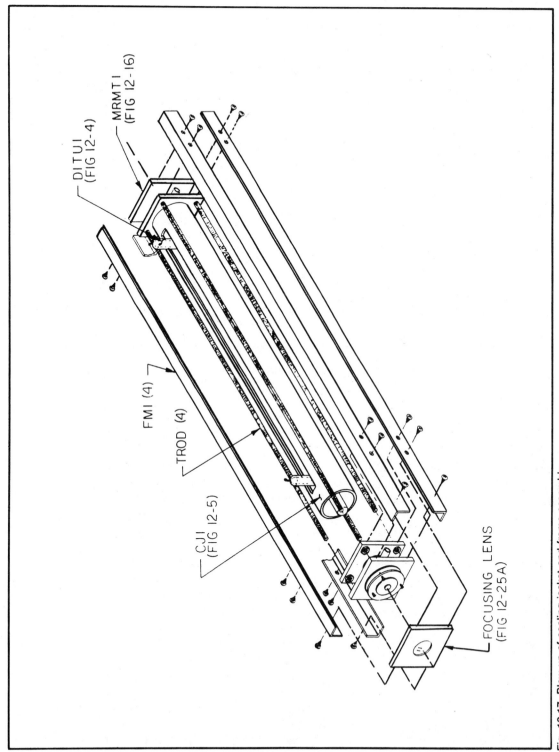

MRMT I
(FIG 12-16)

DITU I
(FIG 12-4)

FM I (4)

TROD (4)

CJ I
(FIG 12-5)

FOCUSING LENS
(FIG 12-25A)

Fig. 12-17. Blow up of cooling jacket and frame assembly.

199

Table 12-2. Power Supply Parts List.

R1,2,3 8	4	100 kΩ 1/4-watt resistor
R4,6,10	3	1-kΩ 1/4-watt resistor
R5	1	1-Ω 1/2-watt resistor
R7,9	2	10-kΩ 1/4-watt resistor
R11	1	330 Ω 1-watt resistor
C1,2	2	1000 µF/25-V electrolytic cap
LA1,2,3 4	4	Neon lamp with lead
D1,2,3 4,6	5	1N4002 lamp 50-V diode
D5	1	1N4007 1-amp 1-kV diode
Z1	1	6-volt zener 1N5234
S1,2	2	10-amp toggle SPST
S3	1	Key switch (nonremovable key in "on" position)
S4,5	2	Pushbutton switch (NO)
**Q1	1	D40D5 NPN power tap
**T1	1	12-V 100-mA transformer
**T2	1	12,000-volt/60-mA transformer
**VA1	1	Variac 115-V ac/5-A
K1	1	12-V DPDT relay
**K2	1	10-A DPST contactor
CO1	1	3-wire heavy power cord
J1,2	1	15-A ac dual receptacle
J3	1	ac power jack for external "laser warning light"
J4	1	Remote 4-screw barrier strip
PB1	1	5" × 3" .1" × .1" perfboard
M1	1	ac ampmeter 0-10 amps not shown on pictorials
FP1	1	Front panel section Fig. 12-21 use 24 ga. galvanized
RP1	1	Rear panel section Fig. 12-22 use 24 ga. galvanized
TC1	1	Top cover 5/8" plywood
BC1	1	Bottom cover 5/8" plywood

Before K1 can be energized it is necessary for capacitor (C2) to charge to a voltage above that of zener diode (Z1). This is accomplished by the time constant of C2 and resistor R7. Diode D6 provides a safety margin by raising the base-to-emitter conduction level of control transistor Q1 to almost 1 volt. Holding current for Q1 is obtained through resistor R10. You will note transformer (T1) is necessary to supply the low voltage to the control circuits. This low voltage is rectified by diode bridge D1-D4 and filter capacitor C1. System control consists of two optional ac jacks, complete with control switches (S1, S2) and associated pilot lamps and limiting resistors LA1, LA2, R1, R2, respectively. These optional jacks are intended for the vacuum system, vacuum gauge, or other ac powered equipment. Main laser power is controlled via key switch S3. Pilot lamp LA3 ignites with this function indicating that the system is "on." A heavy three-wire power cord with ground feeds the system.

Please note that there are several features incorporated into the system required for BRH compliance. These must be verified for safe operation and consist of the following:

1. Key switch S3 with removable key only in "off" position.

2. Time delay for primary contactor K2 actuation after the key is turned "on" consisting of the charge time of C2 through R7 reaching a voltage higher than zener

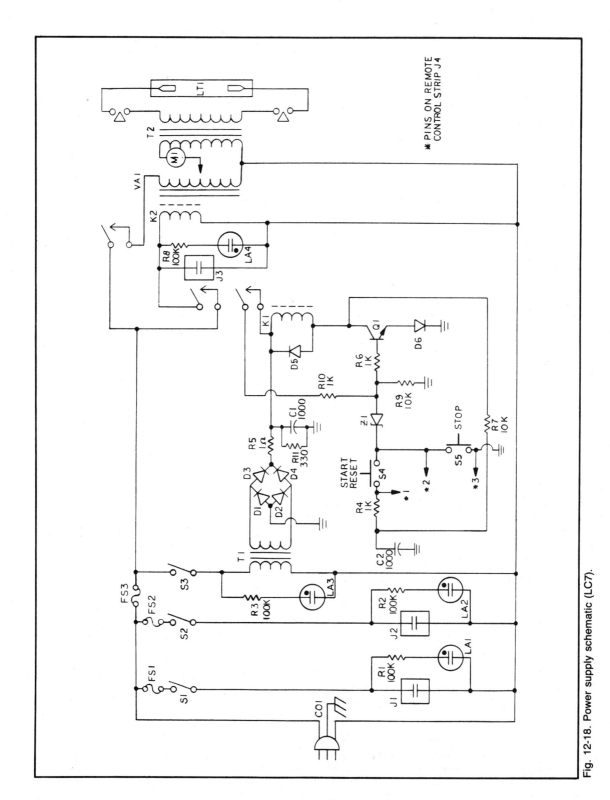

Fig. 12-18. Power supply schematic (LC7).

201

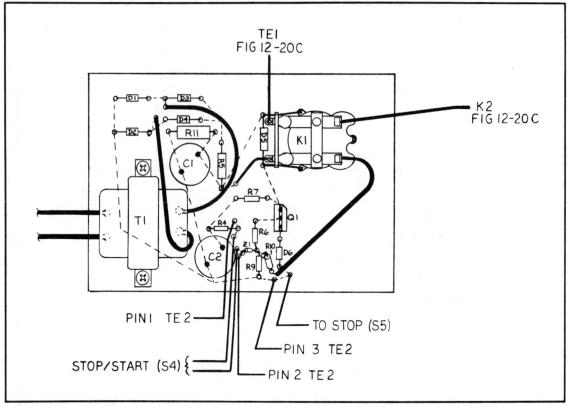

Fig. 12-19. Card assembly.

diode Z1. This time is approximately equal to 10 seconds.

3. Remote control Start/Reset and Stop functions are all obtainable via jack J4.

4. Pilot lamp LA3 gives an indication of system "on" when key switch S3 is energized.

5. Reset operation is required if a power failure occurs. This is accomplished when the holding contact of K1 breaks the circuit and the delay time function now repeats itself only allowing a reset after this time.

6. Interlock as discussed.

STEP 1. Construction of your laser power supply utilizes normal power wiring procedures similar to those used for building ham radio equipment. An excellent source of information may be obtained from the *Radio Amateur's Handbook* published by the ARRL.

STEP 2. Assemble control and delay board as shown in Fig. 12-20. Note polarity of components and avoid wire bridges.

STEP 3. Assembly is shown utilizing fabricated #24 galvanized sheet metal. See layout of front panel (Fig. 12-21) and rear panel (Fig. 12-22). The top and bottom are a finished piece of 5/8" to 3/4" plywood. Other approaches may be considered for the enclosure depending on the builder's requirements. The important thing is to make sure the metal panels are at earth ground via the power cord.

STEP 4. Assemble and wire as shown in wiring aids (Fig. 12-20(A), (B), and (C)). Note the use of #18 wire for primary circuit conductors. These are the leads carrying power to the auxiliary ac jacks J1 and J2 and to variac VA1/transformer T2.

STEP 5. Verify all wiring and check for shorts and other errors.

STEP 6 Turn VA1 fully ccw (off) and all switches to "off."

STEP 7. Short the output terminals of T2 with a secure clip lead. **Danger: The output at these terminals is 120 mA to ground.**

STEP 8. You may wish to connect a ballast such as a 500-watt lamp, toaster, or other high wattage device at this point, as errors will blow fuses and circuit breakers. A variable 115-Vac variac is another alternative.

STEP 9. Check fuses for proper values.

STEP 10. Plug unit into 115-Vac line. No current should flow.

STEP 11. Connect a test lamp (60 watt) to AUX1 and check action of appropriate switch and indicator lamp. Repeat for AUX2. Remove appropriate fuses to check for proper wiring as noted by circuit breaking.

STEP 12. Connect a voltmeter between #1 and #3 of remote control terminal strip (TE2). Turn key "on" and note voltage slowly climbing to 10 volts before start

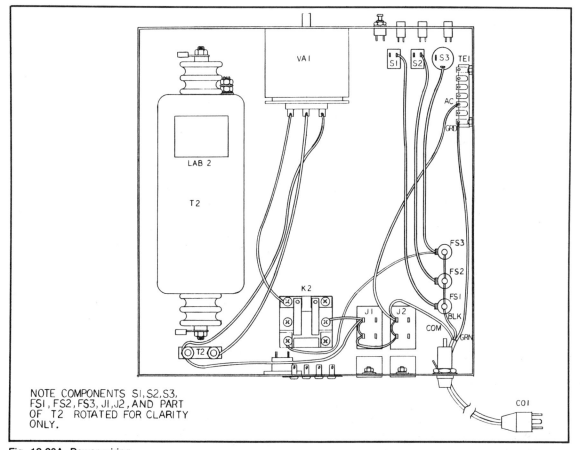

NOTE COMPONENTS S1,S2,S3, FS1,FS2,FS3, J1,J2, AND PART OF T2 ROTATED FOR CLARITY ONLY.

Fig. 12-20A. Power wiring.

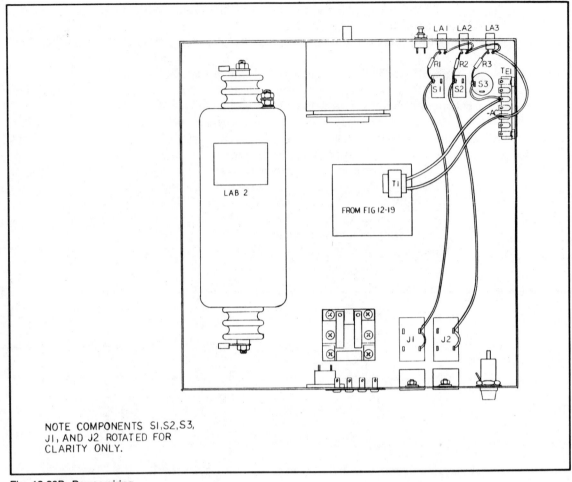

NOTE COMPONENTS S1,S2,S3, J1, AND J2 ROTATED FOR CLARITY ONLY.

Fig. 12-20B. Power wiring.

button S4 can energize relay K2. Note the voltage now dropping back to zero when K2 is energized. Push the stop button and note K2 deenergizing and voltage again climbing to 10 volts before system can be reenergized. This verifies the time delay necessary for BRH compliance.

STEP 13. Remove power cord and reconnect. Note voltage climbing and system requiring a restart with delay. This checks the BRH requirement necessary for the system to require resetting once primary power is lost.

STEP 14. Check action of indication lamp LA3 when key is turned on. Check action of remote jack (J3) for warning lights when K2 is energized. This is an extra for those who wish to utilize the system entry warning lights when the laser is in operation.

STEP 15. Energize system via start switch and slowly turn variac VA1 up to full power. A line current of 5 amps should occur. For those without a proper meter it is suggested to wait until powering up the laser or you may carefully check for a HV arc between each output terminal and ground with a *well-insulated* screwdriver.

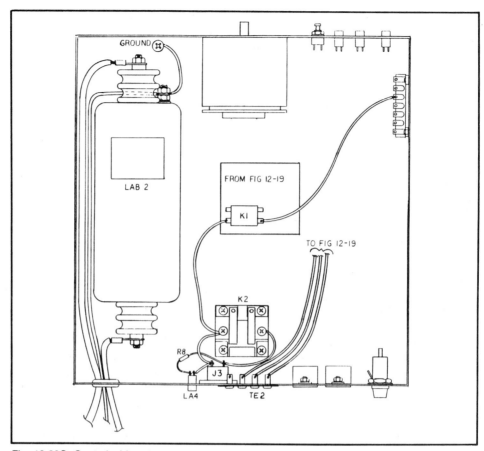

Fig. 12-20C. Control wiring.

STEP 16. Push stop button S5 and deenergize K2. Momentarily make contact to pins 1 and 2 of remote control TE2 and note K2 energizing. Repeat between pin 2 and 3 and note K2 deenergizing. This step checks the action of the remote control function necessary for BRH compliance.

STEP 17. Unit is now ready to use with the laser head. Check for proper labelling and dress of HV wires. Remove short across T2.

The power supply shown produces a variable 0-9 kVac at 60 mA controlled by variac VA1. It is this electrical output that supplies energy to the system by exciting the nitrogen gas in the plasma discharge tube. This ability to control system power will be an obvious advantage when the operator realizes the many combinations of gas pressure and plasma tube current to be experimented with for obtaining either maximum or most efficient output. *These states do not occur simultaneously.*

Metering of the supply is done via meter (M1) that reads transformer (T1) primary current. This approach eliminates the high voltage insulation and mounting requirement involved in placing a current meter in the HV circuit and measuring discharge tube current directly. However, the direct method may be required for certain applications.

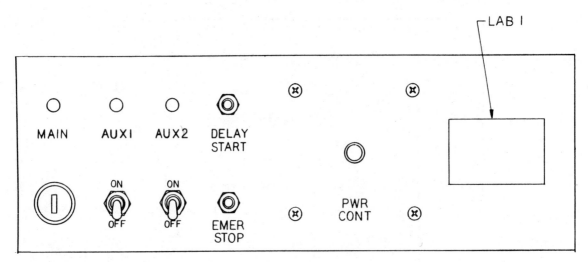

Fig. 12-21. Front panel.

MIRROR ALIGNMENT

A. Position laser head in reference to a helium-neon laser as shown in Fig. 12-23. He-Ne laser should be mounted on a photographic tripod for ease in aligning. Bore-sight laser to discharge tube as shown. This can be achieved by placing a piece of thin paper over the "reflecting" end along with a piece of paper with a small hole in the center over the "exit" end. Laser beam now enters center of "exit" end and is aligned with center of reflecting end. This assures a true bore-sight between the laser head and the He-Ne laser. Firmly secure in this position.

B. Very carefully install reflecting mirror (MIRRI) and seal with "O" ring and vacuum grease.

C. Note reflection from MIRRI of He-Ne laser beam and adjust screws so that it reflects onto back of paper at "exit" end of laser head. It is a good idea to "mike" the position of the mirror adjust washer ADWASI using verniers and noting distance between surfaces of LAWASI. Make a record of these measurements adjacent to the adjust screws. If this is not done it may be difficult to obtain this bore sight when the mirrors are permanently in place.

D. Install exit mirror MIRR1 using same procedure as reflection mirror. Note that the maximum reflective side is in the cavity. This is usually indicated by the symbol >. Adjust the mirror for all back reflections to enter back into the He-Ne laser output aperture. Make sure entering beam spot is still centered as this is the bore axis. Read-just screws for all back reflections to be incident at the He-Ne laser aperture. This adjustment should allow lasing to take place when the system is powered up. Note setting being taken with verniers as done with reflection mirror in previous step.

In the event that mirror adjustment becomes grossly misaligned, the following procedure is suggested without mirror removal.

The reflection mirror adjustment *when exit mirror is in place* requires that the He-Ne laser be properly positioned so that it is as axially coincidental to the bore as possible. This requires a visual sighting using an external sight line or other means and may

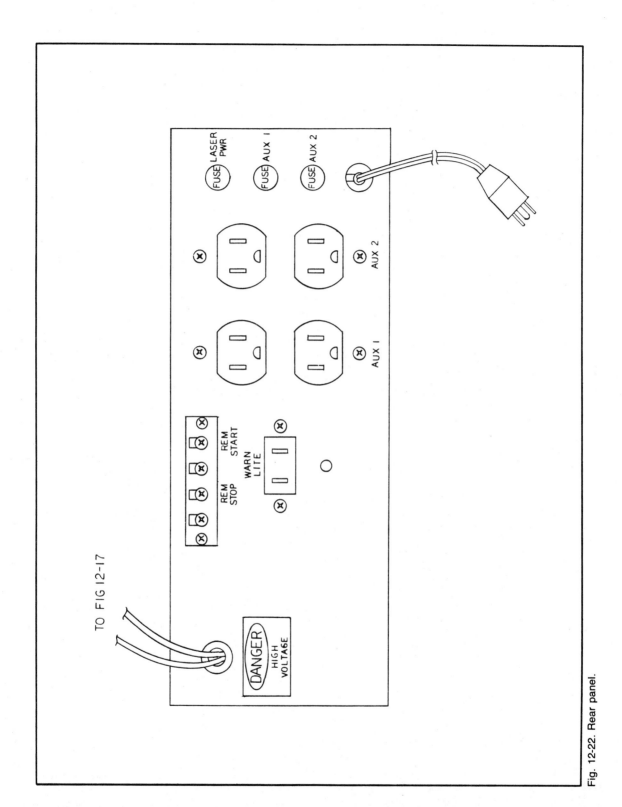

Fig. 12-22. Rear panel.

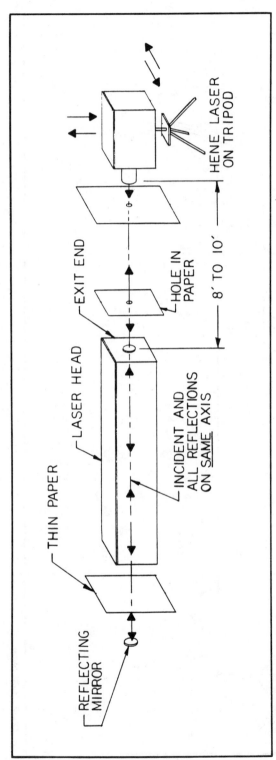

Fig. 12-23A. Alignment of mirrors.

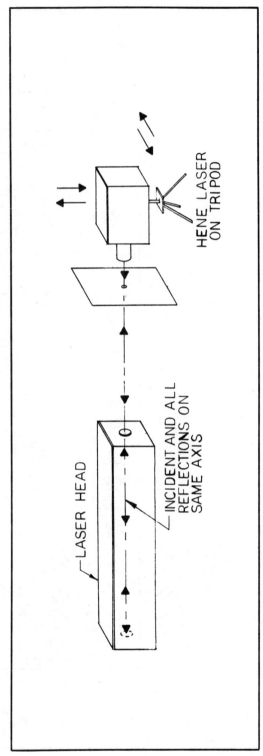

Fig. 12-23B. Adjustment with mirrors in place.

require trial and error *as the inner bore is not accessible at the reflecting end* to actually see the beam location. The reflecting mirror should now be adjusted by eye or with the above vernier measurement for approximate return reflection on the bore axis. The exit mirror may be in any reasonable position.

Place a piece of white paper with a hole for the aperture around the He-Ne laser as shown in Fig. 12-23. There should be a reflection of the beam somewhere on the paper or thereabouts. Be careful not to confuse this reflection as due to the output exit mirror. Proper beam identification can be verified by turning one of the exit mirror adjust screws and noting no change in location. The reflection wanted is the one that the reflecting mirror adjust screw will change and hopefully this should be in rough alignment with the approximated eyeballed adjustments as described. If this reflection cannot be obtained it will be necessary to carefully move the He-Ne laser on the laser head (whichever is easier) very carefully both horizontally and vertically and search for the reflected spot, attempting to get it as near bore-sight as possible. Note that you may obtain reflections of light from the sides of the discharge tube when gross misalignment exists.

When the reflection is located it should be sharp and clear and not contain erroneous reflections. Adjust reflection to the center of the He-Ne laser output aperture. This is the beam reflected back on itself by the reflecting mirror. The exit mirror is also now adjusted to reflect back to the aperture of the He-Ne laser. Alignment should be with all reflected beam spots centered in the exit mirror and all reflections incident on the middle of the He-Ne laser aperture. This adjustment of mirrors should allow lasing to take place with final touch-up being an indication of output power. Note that the reflecting mirror appears to require the most touch-up when initially setting up the system. See Table 12-3.

FINAL ASSEMBLY

STEP 1. Fabricate cover PC1 as shown in Fig. 12-24. Note that if cover is removed the discharge tube is automatically deenergized by the action of the alligator clip or lose crimp lugs disconnecting from EL1, 2. This by no means is to be considered as a safe method of cover removal as a high voltage hazard will exist. Further safety may be built in by electrically connecting PC1 to grounded frame via a flexible piece of wire or braid.

Table 12-3. Required Support Equipment.

VACPMP	Any vacuum pump capable of pumping to a torr or less #1400 2-stage or #1399 1-stage Sargent Welch.
VGAUG	0-100 torr (mm) capsule gauge Leybold Heraus #16068T.
REGUL/FLWVAL	TSA752 2-stage regulator and flow meter from Linde Gas 40-440 ccm.
GASCYL	CO_2 laser mix 9.5% CO_2; 13.5% N_2 from Linde Gas.
**HNE10	.5 mW He-Ne laser for alignment.

**Parts and assemblies individually available from Information Unlimited, P. O. Box 716, Amherst, NH 03031. Write or call (603) 673-4730. Complete kits and finished assemblies are also available.

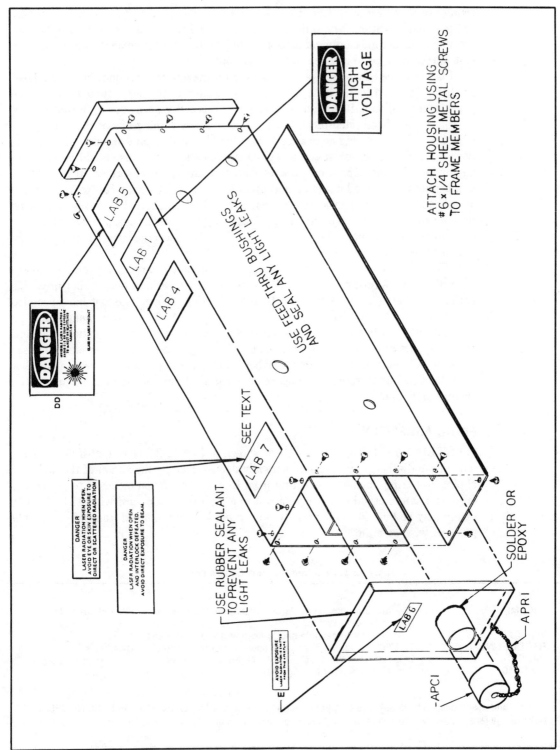

Fig. 12-24. Protective cover and labelling.

STEP 2. Attach appropriate warning labels before using device.

Please note that this laser is a radiation hazard and has the capability of burning and cutting. Obviously, it can cause severe burns to human tissue and is an optical hazard.

Proper posting of the area that the device is used with a warning to personnel is suggested. A special room with warning lights should be allocated for use in experimenting and using the laser with all personnel present wearing protective goggles. The power supply as shown has provisions for powering special warning lights upon system initiation.

The system is shown set up with the laser head and power supply placed on a table. The vacuum pump and gas cylinders are placed on the floor to avoid vibration from the pump motor that could cause beam jitter or modulation. The setup shown is only a suggestion and the user/builder may have his own ideas.

STEP 3. Hook up water supply and allow to fill. (Note for limited use of the laser, a running water supply may not be needed.) Fill water jacket so that discharge tube is totally immersed. Plug input and output hole with a cork stopper to prevent leakage. If running water is required use tygon tubing and limit water pressure as cooling jacket is not made to take pressure but just to provide adequate flowing for cooling.

STEP 4. Connect vacuum lines as shown in Fig. 12-25. Use vacuum grease and hose clamps on all fittings. Attach regulator (REGUL) and flow meter (FLWVAL) to gas tank (GASCYL) and check as shown on instructions that are included with regulator valve. **Make sure regulator is set to less than 5 pounds or gas may blow up this system from over pressure. Turn "off" flow meter.** This must be verified before connecting to laser. Note that a one gallon pickle jar or equivalent filled with cotton is suggested to be connected after the vacuum pump to prevent any oil from getting onto the internal optics of the system. Use a metal cover and solder the input tubes so that it extends into the cotton. The exit tube may just enter beyond the inner surface of the cap. This allows the flow of air through the cotton.

STEP 5. Set up the vacuum gauge as shown on the instructions with the Gilmont vacuum gauge. You may want to make a wooden stand with hose strain relieved. A 0-100 torr *"capsule gauge"* may be preferred for this part because it is easier to handle and not so fragile.

STEP 6. Turn on vacuum pump and allow to pump down for 15 minutes. Note gauge reading less than 1 torr (1 mm). Turn off valve to pump and note vacuum holding. If not, check for faults using conventional leak detection methods. Pinching "off" hoses at various stages sometimes helps to locate the leaking section. Our system pumps down easily to 50 millitorrs in 10 minutes. In the event that there are any leaks in the system they must be fixed for proper operation. There is no quick and dirty method for correcting leaks in a vacuum system. Careful workmanship in constructing and hookup is the best approach.

STEP 7. With system pumped down slowly open valve on gas tank. **Caution! Caution! Again regulator must be set at less than 5 pounds and the flow meter set to "off."** Remember if you allow gas pressure to build up faster than the pump can take it down, a positive pressure will eventually build up to that which the regulator is set. This could blow out the mirrors from their mounts if it is allowed to exceed 5 pounds.

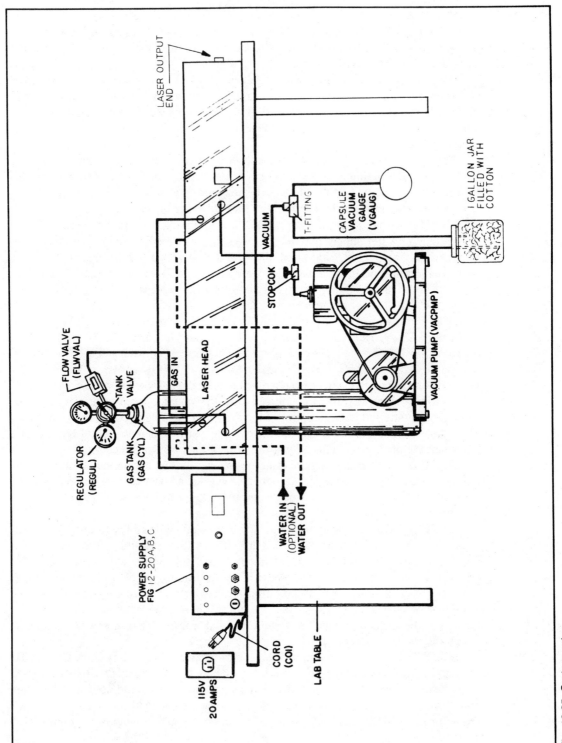

Fig. 12-25. System setup.

STEP 8. Carefully admit gas into the system via the flow meter needle valve. Observe the vacuum gauge and adjust to 15-20 torrs. Allow to stabilize. Note that the pump will take on a different sound as gas is admitted to the system.

STEP 9. **At this point it is necessary to obtain safety glasses.** We use the plastic protection face masks used in shops, etc. Place a piece of wood several inches from the exit mirror. Turn on power and slowly adjust variac (VA1) until tube glows a soft pinkish, purplish. If you're lucky and did all your homework (mirror alignment in head section) the piece of wood should immediately start smoldering with a spot the size of about 1/4″ or 5 to 7 mm. Adjust beam current for 5 amps input to T1 or if reading "direct tube current to about 50 mA."

If output is not detected by the charring of wood, obtain some carbon paper and check for effect. Very carefully tweak the top adjust screw of the reflecting mirror and note carbon paper discoloring or burning. The trick at this point is to carefully tweak the mirrors for maximum output as indicated by the burning effect of the beam. Experience will show that the reflecting mirror is usually the one requiring more frequent touching up as the system is used.

Once mirror settings are optimized it is suggested that various gas pressures and combinations of tube current be experimented with to determine the best parameters for output power. Keep a chart for further reference as each system will be a little bit different. Results will vary but parameters should be close to those specified.

The laser system is capable of projecting a beam of energy across a good size room and immediately burning a hole in whatever it touches. It is very hazardous and should always be terminated into a block of wood as a misdirected beam can start fires and seriously burn flesh. As an example, I was accidentally burned on my upper leg when attempting to reflect the beam onto a target. While I was standing behind the device, part of the beam missed the mirror and got me. Before I knew what happened my pants were burnt and I received a painful burn that eventually blistered and provided an excellent lesson in safety.

APPLICATIONS

In order for your laser to do useful work such as drilling, cutting, etc., it is usually necessary to focus or shape the beam depending on the work required. Lenses and optics for the wavelength output of these lasers are expensive and relatively specialized and we leave the final decision to the user when selecting the optics for the particular laser function. We have listed several suppliers of optical components for CO_2 laser systems. A means for mounting a basic lens system is built into the laser head output end and consists of the extended frame members and a PVC mounting block that easily fits into place. It is adjusted by the sliding action between it and the frame members. Use ingenuity in actual mounting of the lens. **Do not touch the faces.**

The laser is capable of cutting plastic, cloth fabric, cutting styrofoam, etching wood, and many other applications where moderate power is required. Beam direction and position can also be controlled by using a second reflecting mirror and moving it for beam positioning. This sometimes is feasible as the laser head and work piece may now be stationary with the mirror moved for beam placement on the work piece. Use care as most metals will reflect a large percentage of the beam energy.

A simple suggested lens system (Fig. 12-26) is the use of a gallium-arsenide miniscus

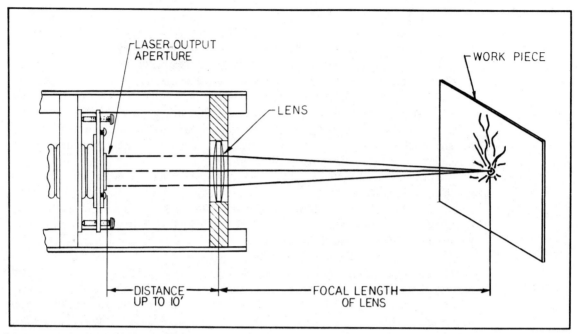

Fig. 12-26. Adjusting the laser's focus.

lens. This lens is available from Information Unlimited and comes in several different focal lengths and diameters. Spot size will not be as small as that obtained by first expanding the beam and then focusing down, however, this method is very expensive.

CHECKLIST REGARDING SAFETY
EQUIPMENT AS REQUIRED FOR BRH COMPLIANCE

1. **Check labels and positions.**
2. **Check action of interlocks.**
3. **Check action of key switch.**
4. **Check indicator and time delay to HV contactor.**
5. **Check reset action.**
6. **Check remote controls.**
7. **Check secureness of protective housing.**
8. **Check aperture cap.**
9. **Check that you and others on premises are using full protective eye wear. Place warning signs at entrance points where laser is being used.**

Chapter 13

HIGH VOLTAGE

Plasma

Tornado Display Lamp/

Sculpture Lamp (PTG1)

P LASMA LAMPS (SOMETIMES CALLED SCULPTURE LAMPS) PRODUCE BEAUTIFUL and mystifying displays of this fourth state of matter. Long snake-like and colorful finger-like streamers emanate from the lamp's center and create a splashing against the inner walls of the enclosure. This splashing, now of a different color, further adds to the spectacular and pleasing visual display of light and color. These devices usually cost upwards of $500.00 or more.

These plans will allow you to construct the necessary power supply to operate any of the lamps from 6″ to 12″ in diameter. The individual lamps are available from Information Unlimited and come in three sizes and a variety of colors. Cost is well under what you would pay if purchased elsewhere. It is suggested you write or call and request further information from Information Unlimited, P.O. Box 716, Amherst, NH 03031 or phone (603) 673-6493.

Plasma is often considered to be the fourth state of matter. It consists of atoms that are ionized and demonstrates peculiar effects unlike the other three forms of matter.

The device described demonstrates a plasma produced by high-frequency, high-voltage electrical discharge through a low-pressure gas. The plasma created produces a visible and bizarre lighting effect that is totally different from any other lighting phenomenon. Columns of pinkish and purplish plasma are attracted to external influences such as fingers and other objects when placed on or near the display container. These columns of plasma light span the entire length of the display container dancing and writhing with a tornado type effect. Balls of plasma and fingers are created and controlled by simply touching the container. This effect cannot be effectively or justifiably described in words and can only be appreciated when actually observed.

The described device is intended for display purposes such as advertising, conversation pieces, novelty decorations, and special effects, etc. It can also be an educational science-fair project demonstrating plasma controlled electrically and magnetically. Special materials treated by a controlled plasma beam can also be realized.

The device consists of a low-powered high-frequency high voltage, producing the necessary parameters for obtaining the described plasma effect. This generator utilizes conventional electrified circuitry consisting of a transistor switching the ferrite core of a high-voltage resonant transformer (similar to a TV flyback). Power for the transistors is obtained from a simple step-down transformer and rectifier combination.

A clear container of suitable size for the display is evacuated to a low pressure of less than 1 torr and includes a single internal discharge electrode similar to that used for neon tubes. The metal cover of a 1 gallon clear glass pickle jar provides an excellent low cost approach to constructing a home made or lab model without expensive glass blowing facilities. The display container must have provision for depressuring and then being permanently sealed. Again this metal cover of a pickle jar makes an excellent choice as a piece of copper capillary can be directly soldered to it forming a good vacuum tight seal and allowing pinching off for sealing. Should the display container require repump down the pinched capillary is opened for reconnection to the vacuum system. The display container is mounted on a suitable stand that houses the generator beneath it. The entire assembly resembles a water cooler.

The following instructions show how to construct a device capable of generating a "plasma tornado." This phenomenon utilizes nature's fourth state of matter to produce this effect. While the device doesn't do anything really useful, with the exception of deodorizing putrid air, it does demonstrate an interesting display of this form of energy. Several local pubs in the area have purchased these units assembled and ready to use. They place them on the bar or other appropriate location and allow the customers to control the plasma tornado using their fingers and hands etc.

The display is inside of a glass enclosure and resembles a tornado shape of glowing and swirling plasma. It dances and jumps to anything brought near it and is highly visible even in normal fluorescent lighting. This sensitivity to any external capacity creates many bizarre and seemingly striking effects. The plasma also can light up a fluorescent lamp when brought near the glass enclosure without any wires or connections of any kind. This feature demonstrates the highly radiative properties of the plasma field and provides an excellent Science-Fair Project.

The generator is an extremely interesting conversation piece and is unlike anything else that most people have yet seen. Its theory is very basic but yet it still seems to amaze most people who do not understand it.

THEORY OF OPERATION

An evacuated glass container is sealed and pumped down to 0.5 to 2 torrs of pressure. A metal cap seals the container and serves as an electrode for charging the remaining thin gas mixture. The voltage applied to the cap is at a potential of 10 to 20 thousand volts and is at a high frequency of approximately 25 kHz. The capacitive effect of the thin gas causes current to flow creating the plasma discharges. One may visualize the device in the following manner: a capacitor is formed by the conductive gas inside the container forming one plate, the glass envelope being the insulating dielectric and the outer air serving as the other plate. Any conducting object brought near the container now only enhances this effect and appears to draw the plasma arc to the point of contact.

The vacuum will vary along with the physical parameters of the container and can be adjusted to enhance the type of discharge desired. A pressure where the plasma discharges are most defined may be critical. Increases will create a broken wispy effect where a further decrease will broaden the discharge making it less pronounced. Further variances from the above will eliminate the discharge completely.

The effect of where conduction of a gas peaks at a certain pressure is known as the *Townsend effect* and becomes an important factor in the design of vacuum systems where medium to high voltages are encountered. The device as described does not use any gas other than the existing atmosphere rarefied by evacuation. Other colors and effects are limitless when the builder chooses to charge the unit with other gases or combinations of pressures etc.

CIRCUIT DESCRIPTION

Power is obtained by polarized plug (CO1) and is fused via (FU1) before energizing the primary of (T1). T1 steps down the 115 V power line voltage to 12 volts where diodes (D1, 2, 3, 4) rectify the ac to pulsating dc. This pulsating waveform is smoothed out by capacitor (C1) to 12 Vdc. The collectors of (Q1) and (Q2) alternately switch the 12 V to the primary of resonant transformer (T2) at a period determined by its resonant frequency. See Fig. 13-1 and Table 13-1.

Base drive is obtained through the base winding and emitter return resistor (R2). R1 starts the oscillator. T2 now produces a stepped-up high-frequency voltage determined by the ratio of primary to secondary "Q" of the resonant transformer. The output of (T2) is now connected to the discharge (DE1) electrode being the metal cover of the display container (DIP1).

The highly conductive rarefied atmosphere inside of the container now acts as one plate of a large capacitor. The glass walls of the container serves as the insulating dielectric while the outer air around the container serves as the second plate. The high frequency, high voltage now encounters a discharge path whose impedance is determined by the capacitive reactance of the system. The high frequency current now flows through the inside atmosphere as a plasma and seeks out the least path of impedance, that is, any external object brought near the container. A higher capacitance is formed at the point and attracts the plasma due to the increased capacity effect.

SPECIAL NOTE ON DIFFERENT GASES

The device shown utilizes the spectral and conductive property of rarefied air as a vacuum is drawn down. A variety of effects are possible and can further be enhanced by the admission of other gases at different pressures. The combinations are many with an equal amount of different display phenomena. Suggested gases are helium, neon, carbon dioxide, oxygen, krypton, argon or any combination of these relatively inert gases. **Caution! Stay away from explosive gases and combinations such as hydrogen, fluorine, chlorine, methane, etc. Always use inert gases. Observe all safety precautions when using high-pressure sources and high vacuums. Check with a chemist if in doubt.**

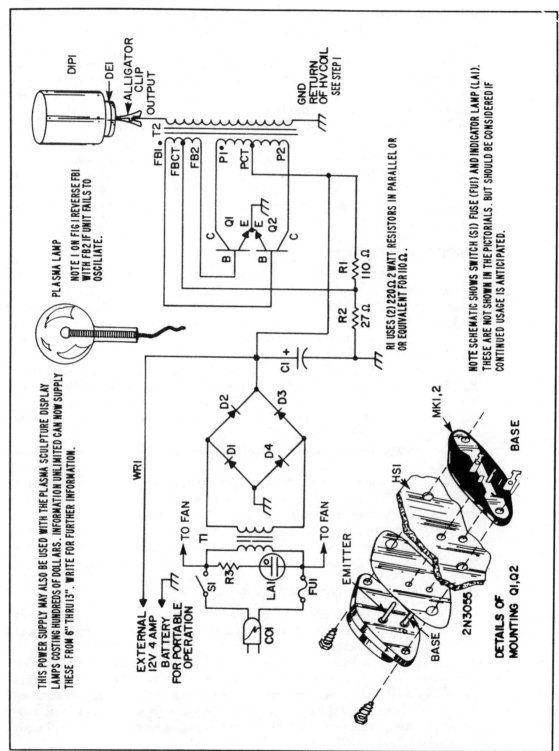

Fig. 13-1. Schematic.

Table 13-1. Plasma Tornado Display Lamp (PTG1) Parts List.

R1	2	Use two 220 Ω resistors in parallel for 110 Ω
R2	1	27 Ω 1/2-watt resistor
R3	1	100 kΩ 1/4-watt resistor (For lamp LA1 not shown)
C1	1	8000 μF/16 V elect cap
D1,2,3,4	4	1N5401 3-amp pwr rectifier
Q1,2	2	2N3055 NPN TO3 transistors
FU1	1	1-A fuse (Not shown in pictures)
CO1	1	2 wire pwr cord
T1	1	12 V 3-A pwr transformer
*T2	1	Ferrite TV flyback transformer Fig. 13-2
WR1	24″	#20 vinyl wire
WR2	24″	#18 vinyl wire
MK1,2	2	TO3 mtg kits
TE1	1	7-lug terminal strip
HS1	1	Dual TO3 heat sink
CL1	1	Alligator clip
BK1	1	Special fabricated bracket #24 galvanized or #22 aluminum (Fabbed per Fig. 13-3)
EN1	1	Large plastic container reworked plant pot etc.
BU3	1	1/2″ bushing for wires of T2
BU2	1	Bushing for line cord
*CV1/JAR1	1	1 gallon jar with metal cover (See text Fig. 13-6)
SW1/NU1		6-32 × 3/8″ screws/nuts
WN1,2	2	Small wire nuts
SW2	2	6-32 × 1″ screws
SW3	3	#6 × 1/2″ self tapping for securing EXT1 to BK1
EXT1	1	Extended tube 3-1/2″ OD × 7″ sked 40 PVC
CA1,2,3	3	3-1/2″ plastic caps
*PV1	1	1/8″ brass petcock and fitting (Intended for experimental lab use for easier changing of gases)
CV1	1	Metal cover to jar (reworked) Fig. 13-6.
BU1	1	Small knurled brass nut
COP1	1	1/8″ copper capillary (Use 1/8″ automobile vac hose and pinch off with clamp)
FAN1	1	Rotron or similar fan for continuous use
BR2	1	Brackets for above fan

*Optional 6″, 8″, and 12″ Plasma Sculpture Lamps in several colors available. Write or call Information Unlimited, P.O. Box 716, Amherst, NH 03031 (603) 673-6493 for pricing and availability. A complete kit of the above or an assembled and tested unit is also available. Parts marked with an asterisk are individually available.

ASSEMBLY STEPS

1. Layout and identify all parts and pieces.

2. Rework flyback transformer T2 as shown in Fig. 13-2. Determine ground return of output winding by selecting the pin or lead with the highest resistance reading between it and the output lead.

3. Fabricate mounting bracket BK1 from a piece of 22 gauge aluminum or galvanized sheet metal. Use the dimensions given in Fig. 13-3. It is a good idea to trial position all of the parts and work the hole location to assure proper fit, location, etc. (See Fig. 13-4.) The bracket supplied with the kit is fabbed as shown for the optional fan.

4. Fabricate extender tube EXT1 from a 7″ length of 3″ schedule 40 PVC (actual 3-1/2″ OD). Use a hole saw to place a 1-3/4″ hole located with its center two inches from the end. This hole is used for access to attach the high voltage/high frequency clip from T2 to the plasma tube and must be as short and direct as possible. See Fig. 13-5.

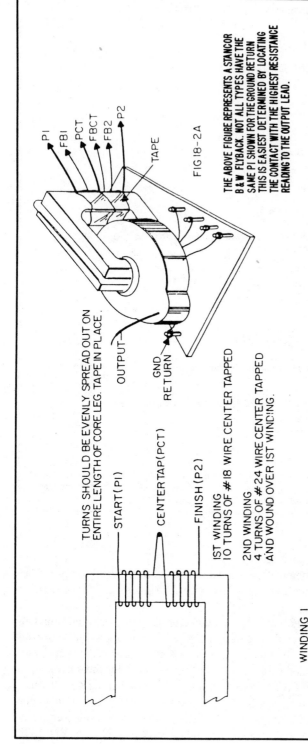

PI
FBI
PCT
FBCT
FB2
P2

TAPE

FIG 18-2A

THE ABOVE FIGURE REPRESENTS A STANCOR B & W FLYBACK. NOT ALL TYPES HAVE THE SAME PI SHOWN FOR THE GROUND RETURN. THIS IS EASIEST DETERMINED BY LOCATING THE CONTACT WITH THE HIGHEST RESISTANCE READING TO THE OUTPUT LEAD.

TURNS SHOULD BE EVENLY SPREAD OUT ON ENTIRE LENGTH OF CORE LEG. TAPE IN PLACE.

START (PI)

OUTPUT

CENTER TAP (PCT)

GND RETURN

FINISH (P2)

1ST WINDING
10 TURNS OF #18 WIRE CENTER TAPPED

2ND WINDING
4 TURNS OF #24 WIRE CENTER TAPPED AND WOUND OVER 1ST WINDING.

WINDING I

TWO NEW WINDINGS ("PRIMARY" AND "FEEDBACK" WINDINGS) ARE ADDED TO THE FLYBACK TRANSFORMER (T2) THAT CONNECTS TO THE DRIVER TRANSISTORS. THESE WINDINGS ARE HAND WOUND ON THE BOTTOM LEG OF THE FERRITE CORE WHERE THE ORIGINAL TWO TURN FILAMENT WINDING WAS LOCATED. REMOVE AND DISCARD THE ORGINAL FILAMENT WINDING. IN ITS PLACE, WIND FIRST A TEN-TURN, CENTER-TAPPED WINDING (DESIGNATED PI-P2) USING APPROXIMATELY 30 INCHES OF #18 OR LARGER INSULATED HOOKUP WIRE. THIS IS EASILY ACCOMPLISHED BY WINDING FIVE TURNS AT ONE END OF THE CORE AND THEN TWISTING A LOOP IN THE FREE END BEFORE ADDING THE SECOND FIVE TURNS. THE COMPLETE TEN TURN WINDING SHOULD THEN BE HELD IN PLACE WITH A TURN OR TWO OF ELECTRICAL TAPE WITH THE TWO ENDS (PI & P2) AND THE CENTER TAP (PCT) LOOP ALL PROTRUDING. CONNECTION CAN BE MADE TO THE CENTER TAP LOOP WHEN THE INSULATION HAS BEEN CAREFULLY REMOVED. IF IT BECOMES NECESSARY TO CUT THE CENTER TAP LOOP, BE SURE THAT THE TWO ENDS ARE SCRAPED AND JOINED TO FORM A MECHANICAL AS WELL AS AN ELECTRICAL CENTER TAP CONNECTION TO THE WINDING.

WINDING II

THE SECOND WINDING (FEEDBACK) SHOULD BE WOUND DIRECTLY ON TOP OF THE FIRST, BUT IT SHOULD ONLY HAVE A TOTAL OF FOUR TURNS—TWO EACH SIDE OF THE CENTER TAP. WIND TWO TURNS OF #22 HOOK-UP WIRE, PULL AND TWIST A CENTER TAP LOOP (FBCT) AND WIND THE OTHER TWO TURNS. TAPE THIS WINDING IN PLACE ON TOP OF THE FIRST. DO NOT LET THE CENTER TAP LOOPS OF THE TWO WINDINGS TOUCH EACH OTHER

Fig. 13-2. T2 winding instructions.

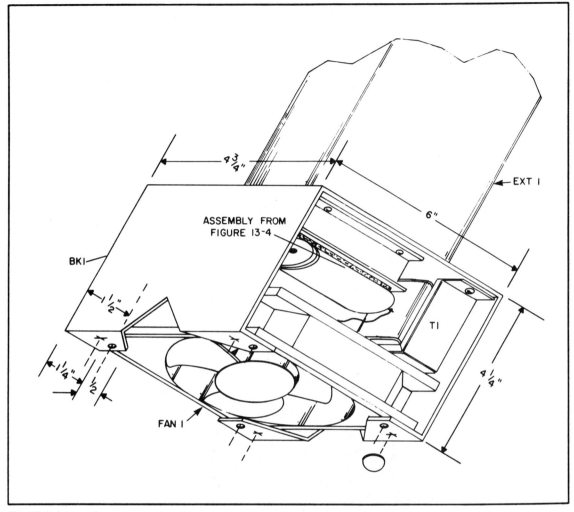

Fig. 13-3. Lower view.

5. Assemble transformer (T1) to bracket (BK1) via screws/nuts (SW1/NU1). Fasten terminal strip TE1 to heat sink HS1 via long screws and nuts (SW2/NU1). It may be necessary to drill these holes in heat sink (between 1st and 2nd fin). See Fig. 13-4. Attach Q1, Q2 to HS1 as shown by inset in Fig. 13-1. Note to remove any burrs that might puncture through the mica insulating washer. Note the location of the base "B" and emitters "E" of these transistors. Wire as shown and connect wires from T1 and T2. Note identification wires for T2 using Figs. 13-1, 13-2, and 13-4. These wires are routed through busings BU2 and should just be long enough to allow positioning of T2 to about 1″ off the surface of BK1. Make final assembly and secure all parts after you have double-checked wiring for accuracy, shorts, bad soldering, etc.

6. Apply power to T1 and note an arc of nearly 1″ being able to be drawn from the output lead of T2. Allow to remain on for several minutes and check for excessive heat. If continuous use is anticipated it will be necessary to cool the heat sink and com-

ponents with a fan as shown in Fig. 13-3. The circuit does have a tendency to run on the hot side due to the switching of Q1 and Q2. See note 1 on Fig. 13-1, if unit fails to oscillate.

7. Obtain a large 1-gallon jar with a metal cover and assemble as shown Figs.

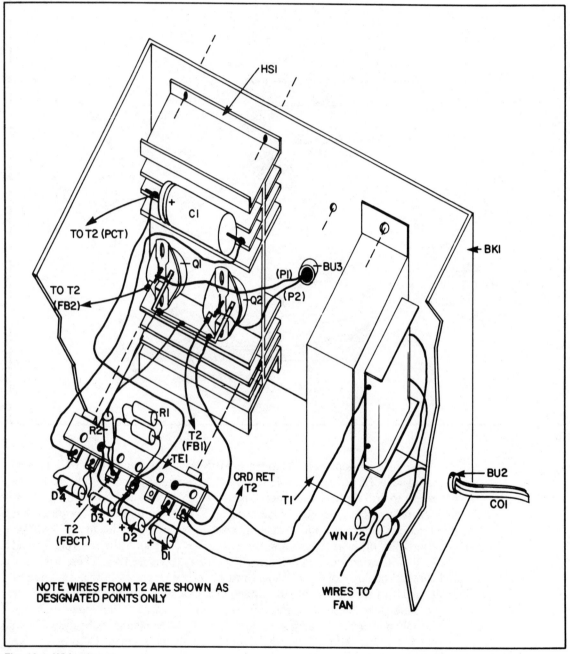

Fig. 13-4. Wiring layout.

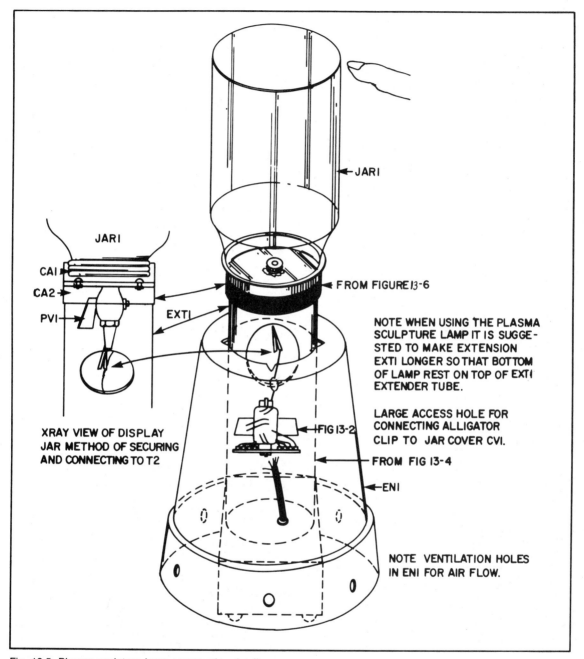

JARI

CAI
CA2
PVI
EXTI

JARI

FROM FIGURE 13-6

NOTE WHEN USING THE PLASMA
SCULPTURE LAMP IT IS SUGGE-
STED TO MAKE EXTENSION
EXTI LONGER SO THAT BOTTOM
OF LAMP REST ON TOP OF EXTI
EXTENDER TUBE.

LARGE ACCESS HOLE FOR
CONNECTING ALLIGATOR
CLIP TO JAR COVER CVI.

FIG 13-2

FROM FIG 13-4

ENI

XRAY VIEW OF DISPLAY
JAR METHOD OF SECURING
AND CONNECTING TO T2

NOTE VENTILATION HOLES
IN ENI FOR AIR FLOW.

Fig. 13-5. Plasma sculpture lamp construction details.

13-6 and 13-7. Note the two approaches shown using a petcock for convenient gas charging or the more permanent capillary approach.

8. Attach two 3-1/2″ plastic caps (CA1, 2) together as shown in the inset of Fig.
13-5. This helps to secure the display jar from toppling over and breaking creating a

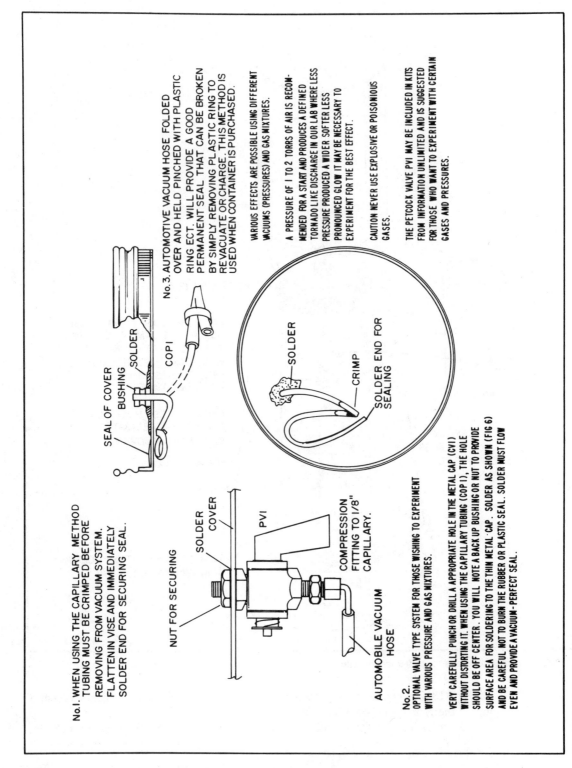

No. 1. WHEN USING THE CAPILLARY METHOD TUBING MUST BE CRIMPED BEFORE REMOVING FROM VACUUM SYSTEM. FLATTEN IN VISE AND IMMEDIATELY SOLDER END FOR SECURING SEAL.

NUT FOR SECURING

SOLDER

COVER

PVI

COMPRESSION FITTING TO 1/8" CAPILLARY.

AUTOMOBILE VACUUM HOSE

No. 2.
OPTIONAL VALVE TYPE SYSTEM FOR THOSE WISHING TO EXPERIMENT WITH VARIOUS PRESSURE AND GAS MIXTURES.

VERY CAREFULLY PUNCH OR DRILL A APPROPRIATE HOLE IN THE METAL CAP (CVI) WITHOUT DISTORTING IT. WHEN USING THE CAPILLARY TUBING (COP I), THE HOLE SHOULD BE OFF CENTER. YOU WILL NOTE A BACK UP BUSHING OR NUT TO PROVIDE SURFACE AREA FOR SOLDERING TO THE THIN METAL CAP. SOLDER AS SHOWN (FIG 6) AND BE CAREFUL NOT TO BURN THE RUBBER OR PLASTIC SEAL. SOLDER MUST FLOW EVEN AND PROVIDE A VACUUM - PERFECT SEAL.

SEAL OF COVER BUSHING

SOLDER

COPI

No. 3. AUTOMOTIVE VACUUM HOSE FOLDED OVER AND HELD PINCHED WITH PLASTIC RING ECT. WILL PROVIDE A GOOD PERMANENT SEAL THAT CAN BE BROKEN BY SIMPLY REMOVING PLASTIC RING TO REVACUATE OR CHARGE. THIS METHOD IS USED WHEN CONTAINER IS PURCHASED.

VARIOUS EFFECTS ARE POSSIBLE USING DIFFERENT VACUUMS (PRESSURES) AND GAS MIXTURES.

A PRESSURE OF 1 TO 2 TORRS OF AIR IS RECOMMENDED FOR A START AND PRODUCES A DEFINED TORNADO LIKE DISCHARGE IN OUR LAB WHERE LESS PRESSURE PRODUCED A WIDER SOFTER LESS PRONOUNCED GLOW IT MAY BE NECESSARY TO EXPERIMENT FOR THE BEST EFFECT.

CAUTION NEVER USE EXPLOSIVE OR POISONIOUS GASES.

THE PETCOCK VALVE PVI MAY BE INCLUDED IN KITS FROM INFORMATION UNLIMITED AND IS SUGGESTED FOR THOSE WHO WANT TO EXPERIMENT WITH CERTAIN GASES AND PRESSURES.

SOLDER

CRIMP

SOLDER END FOR SEALING

Fig. 13-6. Methods of sealing.

224

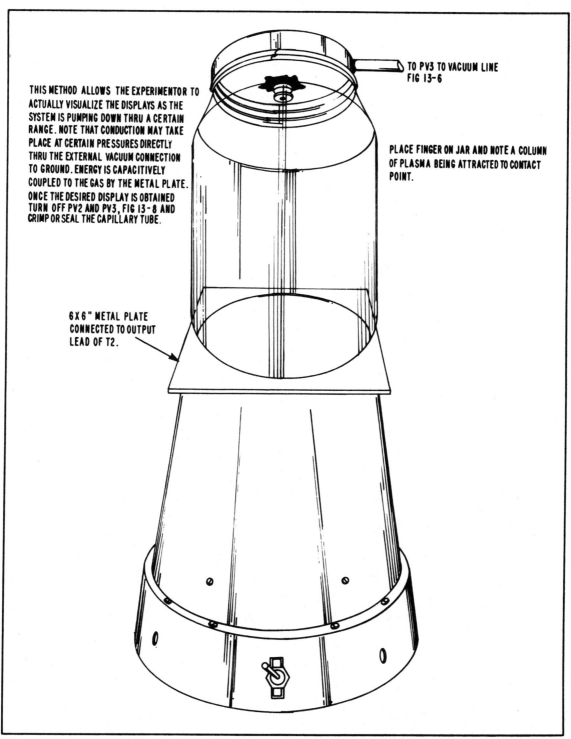

THIS METHOD ALLOWS THE EXPERIMENTOR TO
ACTUALLY VISUALIZE THE DISPLAYS AS THE
SYSTEM IS PUMPING DOWN THRU A CERTAIN
RANGE. NOTE THAT CONDUCTION MAY TAKE
PLACE AT CERTAIN PRESSURES DIRECTLY
THRU THE EXTERNAL VACUUM CONNECTION
TO GROUND. ENERGY IS CAPACITIVELY
COUPLED TO THE GAS BY THE METAL PLATE.
ONCE THE DESIRED DISPLAY IS OBTAINED
TURN OFF PV2 AND PV3, FIG 13-8 AND
CRIMP OR SEAL THE CAPILLARY TUBE.

TO PV3 TO VACUUM LINE
FIG 13-6

PLACE FINGER ON JAR AND NOTE A COLUMN
OF PLASMA BEING ATTRACTED TO CONTACT
POINT.

6 X 6" METAL PLATE
CONNECTED TO OUTPUT
LEAD OF T2.

Fig. 13-7. Sample test method.

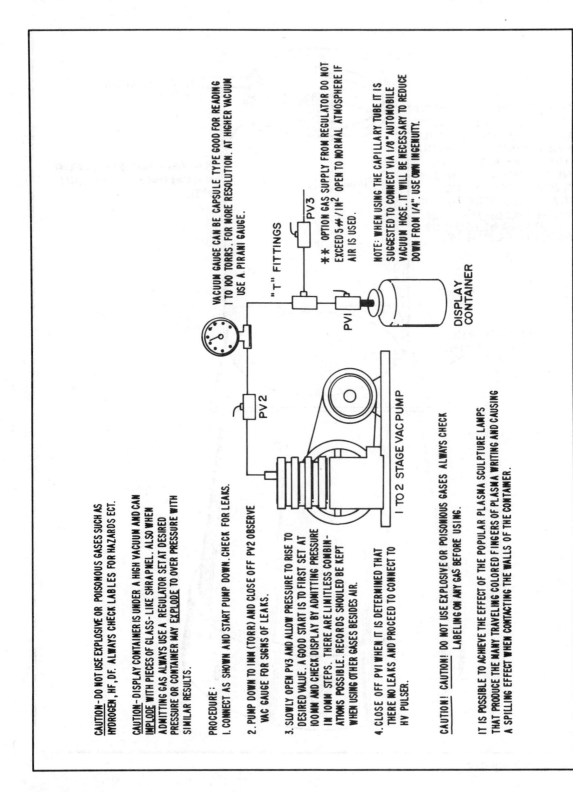

CAUTION – DO NOT USE EXPLOSIVE OR POISONOUS GASES SUCH AS HYDROGEN, HF, DF. ALWAYS CHECK LABLES FOR HAZARDS ECT.

CAUTION – DISPLAY CONTAINER IS UNDER A HIGH VACUUM AND CAN IMPLODE WITH PIECES OF GLASS- LIKE SHRAPNEL. ALSO WHEN ADMITTING GAS ALWAYS USE A REGULATOR SET AT DESIRED PRESSURE OR CONTAINER MAY EXPLODE TO OVER PRESSURE WITH SIMILAR RESULTS.

PROCEDURE:

1. CONNECT AS SHOWN AND START PUMP DOWN. CHECK FOR LEAKS.

2. PUMP DOWN TO 1MM (TORR) AND CLOSE OFF PV2 OBSERVE VAC GAUGE FOR SIGNS OF LEAKS.

3. SLOWLY OPEN PV3 AND ALLOW PRESSURE TO RISE TO DESIRED VALUE. A GOOD START IS TO FIRST SET AT 100MM AND CHECK DISPLAY BY ADMITTING PRESSURE IN 10MM STEPS. THERE ARE LIMITLESS COMBINATIONS POSSIBLE. RECORDS SHOULED BE KEPT WHEN USING OTHER GASES BESIDES AIR.

4. CLOSE OFF PV1 WHEN IT IS DETERMINED THAT THERE NO LEAKS AND PROCEED TO CONNECT TO HV PULSER.

CAUTION! CAUTION! DO NOT USE EXPLOSIVE OR POISONOUS GASES ALWAYS CHECK LABELING ON ANY GAS BEFORE USING.

IT IS POSSIBLE TO ACHIEVE THE EFFECT OF THE POPULAR PLASMA SCULPTURE LAMPS THAT PRODUCE THE MANY TRAVELING COLORED FINGERS OF PLASMA WRITING AND CAUSING A SPILLING EFFECT WHEN CONTACTING THE WALLS OF THE CONTAINER.

VACUUM GAUGE CAN BE CAPSULE TYPE GOOD FOR READING 1 TO 100 TORRS. FOR MORE RESOLUTION. AT HIGHER VACUUM USE A PIRANI GAUGE.

"T" FITTINGS

PV3

** OPTION GAS SUPPLY FROM REGULATOR DO NOT EXCEED 5#/IN² OPEN TO NORMAL ATMOSPHERE IF AIR IS USED.

NOTE: WHEN USING THE CAPILLARY TUBE IT IS SUGGESTED TO CONNECT VIA 1/8" AUTOMOBILE VACUUM HOSE. IT WILL BE NECESSARY TO REDUCE DOWN FROM 1/4". USE OWN INGENUITY.

PV1

DISPLAY CONTAINER

PV2

1 TO 2 STAGE VAC PUMP

Fig. 13-8. Pump down and filling of display tube.

226

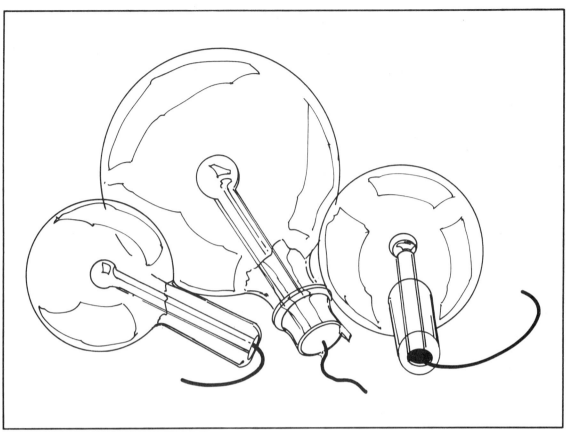

Fig. 13-9. Examples of plasma lamps available thru Information Unlimited - sizes are 6″, 9″, 12″ diameters with different colors.

dangerous hazard. It may be feasible to secure in position with RTV silicon rubber. This is reasonably sturdy and can be disassembled with a minimum of effort. See Fig. 13-6.

9. Perform the sample test method shown in Figs. 13-7 and 13-8. Finish by fabbing EN1 enclosure from a suitable plastic container. Ventilate for air flow. See Fig. 13-9.

10. Demonstrate the device in a darkened room in front of a large mirror. The plasma reflections will make the presentation eerier. Having practiced with the device before the demonstration, you will be able to go through a presentation that will be exciting to those assembled. Have the audience participate. When their interest is at its peak, tell them exactly what you have, and then see if you can get through the pack to play with your Plasma Machine. You may have to practice some crowd-control tactics. If that's your problem, we made your day!

Special Note: It may be necessary to place a small piece of insulated tubing over the control handle of switch S1 if used. Since the operator's body acts as an antenna he will notice a sometimes annoying burning sensation when touching any nearby conductive object.

Chapter 14

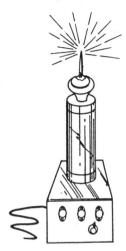

High-Voltage
DC Generator (HVG1)

T HE FOLLOWING DEVICE IS INTENDED AS A LABORATORY SOURCE OF HIGH VOLT-
age direct current that is continuously adjustable from 35,000 to 250,000 volts.
It is not to be mistaken for the low-output-current electrostatic devices such as Van
de Graaff, Wimshurst, Kelvin, or similar generators. The unit will supply a steady current of microamps and easily still support a high voltage. Uses for this type product
range from producing external ion winds to providing the energizing source for small
particle accelerators of special interest in today's Star Wars Defense Program. Devices
of this type are usually used as the energy initializing of particles intended for further
accelerating in linear accelerators, cyclotrons, etc.

**Caution! Caution!—This device produces high voltages that can cause
very painful electrical shocks. Dangerous secondary reactions can cause
physical injury. Attention also must be given to the potential X-ray hazard when certain target metals are exposed to this energy. Never fire a
unit of this type in an explosive or volatile atmosphere as sparking can
cause dangerous ignitions.**

Certain uses of the device are for the demonstration of charged particles and their
ability to travel distances charging up objects, people, an electroscope, etc. In order
to demonstrate this effect, it is necessary to produce voltages of these magnitudes that
may be at a **hazardous shock potential**.

Even though the device only produces small currents it must be treated with caution. Use discretion when using, as it is possible for a person wearing insulated shoes
to accumulate enough of a charge to produce a moderately painful or irritating shock
when he touches a grounded object. The effect while only irritating **could cause in-**

jury to a person in weak physical condition. Quantity of accumulated charge depends on many parameters including humidity, leakage, sizes and types of objects, proximity, etc. Use caution when charging capacitance as a dangerous charge may accumulate.

The device can be utilized in two ways. When the output of the device is terminated into a smooth large surfaced collector it becomes a useful high potential source capable of powering particle accelerators and other related devices. It is well known that high voltage generators usually consist of large smooth surface collectors where leakage is minimized allowing these collective terminals to accumulate high voltage with less current demand. Leakage on the other hand of a high voltage "point" is the result of the repulsion of like charges to the extent that these charges are forced out into the air as ions. The rate of ions produced is a result of the charge density at a certain point. The magnitude of this quantity is a function of voltage and the reciprocal of the angle of projection of the surface. (See Fig. 14-1.) This is why lightning rods are sharply pointed causing the charges to leak off into the air before a voltage can be developed to create the lightning bolt.

It is now evident that to create ions it is necessary to have a high voltage applied to an object such as a needle or other sharp device used as an emitter. Once the ions leave the emitter they possess a certain mobility allowing them to travel moderate distances contacting and charging up other objects by accumulation. It obviously can be built as a producer of negative or positive ions demonstrating a phenomenon that is often regarded as a figure of demerit when building and designing high voltage power supplies. This type of operation also produces a corona with the formation of nitric acid via the production of the ozone produced combining with nitrogen and forming nitrous oxide which, with water, produces this very strong acid. The production of ions also robs the available current from the supply and can be an undesirable effect for many applications.

APPLICATIONS AND EXPERIMENTS USING THE HIGH-VOLTAGE LEAKAGE

1. **St. Elmo's Fire**—This familar discharge glow occurs during periods of high electrical activity. It is a coronal discharge that is brushlike, luminous, and often may be audible, when leaking from charged objects in the atmosphere. It occurs on ship-masts, on aircraft propellers, wings, other projecting parts and on objects projecting from high terrain when the atmosphere is charged and a sufficiently strong electrical

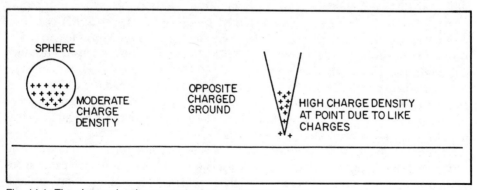

Fig. 14-1. The charge density on a spherical and pointed object.

potential is created between the object and the surrounding air. Aircraft most frequently experience St. Elmo's Fire when flying in or near cumulonimbus clouds, thunder storms, in snow showers, and in dust storms. It is easily artificially produced by using your unit as shown Fig. 14-2.

2. **Flashing Fluorescent Light**—This experiment shows the mobility of the ions and their ability to charge the capacitance in a fluorescent light tube and discharge in the form of a flash (See Fig. 14-3). Perform the following:

☐ Connect a needle or other sharp object to point in a desired direction. Use clay or wax.

☐ Carefully hold a 20-watt fluorescent tube and turn off lights. Allow eyes to become accustomed to total darkness.

☐ Hold end about 3 feet from needle and note lamp flickering. Increase distance and note flicker rate decreasing. Under ideal conditions, and total darkness, the lamp will flicker up to a considerable distance from the source. Use caution in total darkness. Hold lamp by glass envelope and touch end pins to grounded object. **Caution: Remain clear of device as a painful shock with dangerous secondary reactions may result.** Flash time is equivalent to the familiar equation: $T = CV/I$ where T is the time between flashes, V is the flash breakdown voltage, C is the inherent capacity in the tube, and I is equivalent to the amount of ions reaching the lamp and obviously decreases by the 5/2 power of the distance.

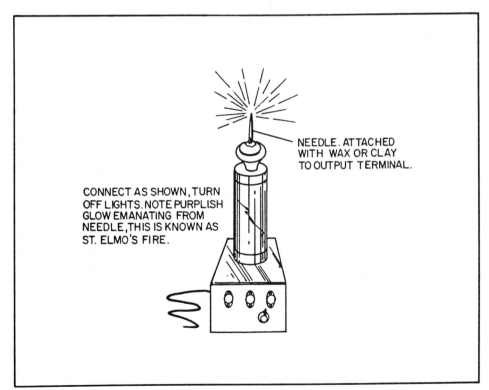

NEEDLE. ATTACHED WITH WAX OR CLAY TO OUTPUT TERMINAL.

CONNECT AS SHOWN, TURN OFF LIGHTS. NOTE PURPLISH GLOW EMANATING FROM NEEDLE, THIS IS KNOWN AS ST. ELMO'S FIRE.

Fig. 14-2. Demonstration of St. Elmo's fire.

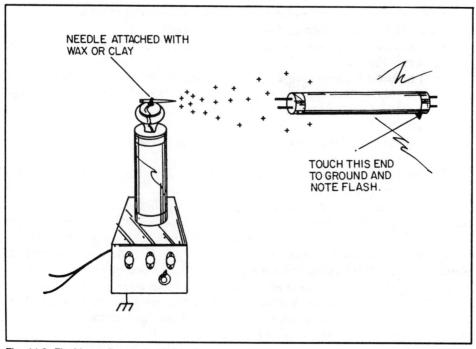

NEEDLE ATTACHED WITH
WAX OR CLAY

TOUCH THIS END
TO GROUND AND
NOTE FLASH.

Fig. 14-3. Flashing a fluorescent light.

3. **Ion Charging**—This demonstrates the same phenomena as in Experiment 2, but in a different way (see Fig. 14-4). Perform the following:

☐ Set up unit as for the Experiment shown in Fig. 14-2.
☐ Set up objects as shown.
☐ Note spark occurring as a result of ion accumulation on sphere. Increase distance and note where spark becomes indistinguishable.
☐ Obtain a subject brave enough to stand a moderate electric shock. **(Use Caution as a person with a heart condition should not be near this experiment.)**
☐ *Have subject stand on an insulating surface* and then touch a grounded or large metal object. Rubber or similar soled shoes will enhance the effect.

4. **Ion Motor**—This dramatically demonstrates Newton's Law of action producing reaction. Escaping ions at high velocity produce reaction. This is a viable means of propulsion for spacecraft where hypervelocities may approach the speed of light in this frictionless environment. Perform as in Fig. 14-5.

☐ Form piece of #18 wire as shown. For maximum results carefully balance and provide minimum friction at pivot point. There are many different methods of performing this experiment with far better results. We leave this to the experimenter bearing in mind that a well made, balanced rotor can achieve amazing rpm.
☐ Note as rotor spins giving off ions, that one's body hair will bristle, nearby objects will spark and a cold feeling will persist from the ion wind produced.

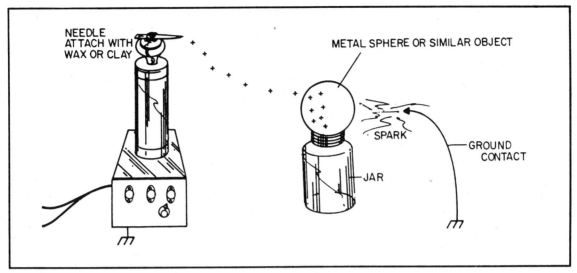

Fig. 14-4. Ion charging experiment.

5. **Demonstrates the Destructive Shocking Power and Penetration of a Spark Discharge**—This property can be used for drilling micro-sized holes in many nonconductive materials. The holes are clean and relatively round and can be seen by holding the sample material up to a bright light. They can be very useful for making pin holes for optical purposes. Perform as in Fig. 14-6. **Use Caution when approaching output terminal. Rig up a well grounded discharge stick. Use ingenuity.**

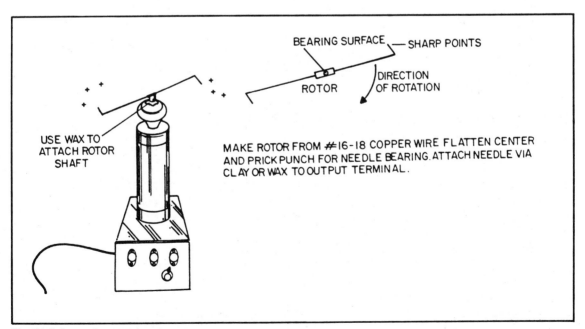

Fig. 14-5. Ion motor experiment.

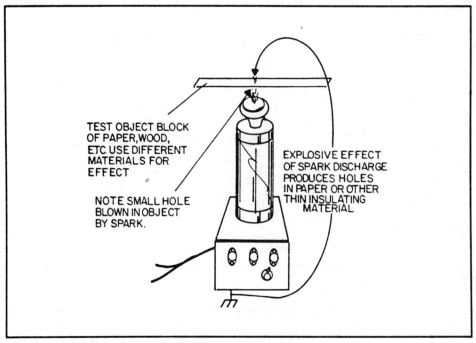

TEST OBJECT BLOCK OF PAPER, WOOD, ETC. USE DIFFERENT MATERIALS FOR EFFECT

NOTE SMALL HOLE BLOWN IN OBJECT BY SPARK.

EXPLOSIVE EFFECT OF SPARK DISCHARGE PRODUCES HOLES IN PAPER OR OTHER THIN INSULATING MATERIAL

Fig. 14-6. Spark discharge experiment.

6. Lightning Generator—Perform as in Fig. 14-7.

☐ Construct two units but reverse all the diodes in the multiplier stages to produce a negative output in the second unit.

☐ Obtain some large smooth metal objects that resemble spheres as nearly as possible. Place them as terminals on units and remove any outward protruding sharp surfaces. Object should be 3″ or more in diameter, use utensils, etc.

☐ Locate units as shown.

7. Experiment Using Your Unit in a Gun Configuration—Capacitor charging performs as in Fig. 14-8.

The unit generates ions that are accumulated on the insulated spherical object charging it theoretically to its open circuit potential (this in practice doesn't occur due to leakage, etc). The object accumulates a voltage equal to $V = IT/C$. Note the unit is also directly grounded to increase this effect by producing the necessary electrical mirror image. The quantity Q (Coulombs) of the charge is equal to CV where C = capacitance of object and V = voltage charged. The energy W (joules) stored is equal to 1/2 CV2. The capacitance can be calculated by approximating the area of the shadow of the object projected directly beneath it and calculating the mean separation (use inches).

The capacitance is now approximately equal to C pF = .25 times the projected area divided by the mean separation in inches.

8. Spark Discharge—Demonstrates the difference in spark length between a high and a low leakage surface (see Fig. 14-9). Perform the following:

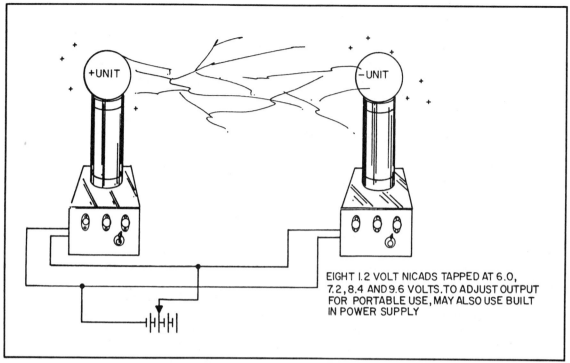

Fig. 14-7. Lightning generator experiment.

Note that the spark may only be 1″ or so with the pointed object while the smooth surface object produces a 3″ to 4″ spark. This is due to the inability of the device to maintain a voltage under the high leakage conditions with the points while the other produces just enough ions to produce a leader for the main discharge.

9. **Demonstrates the Transmission of Energy Via Mobile Ions**—Perform as in Fig. 14-10. Objects are round spheres placed on glass bottles used as insulators. One object is grounded using thin insulated wire. The other object discharges to the

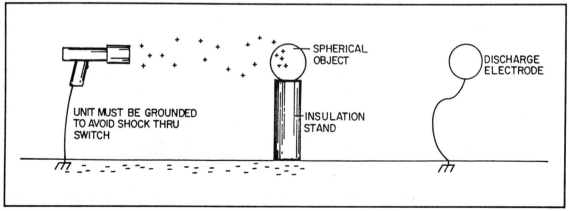

Fig. 14-8. Ion gun experiment.

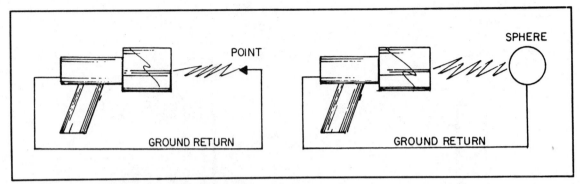

Fig. 14-9. High and low leakage spark discharge experiment.

grounded object. Note grounding wire physically jumping at time of discharge, this phenomena is the result of current producing a mechanical force. As unit is brought closer, length of spark discharge and discharge rate will increase. A discharge length of 1″ is easily obtainable with the unit 4 to 5 feet away. This demonstrates the potential mobile effectiveness of the device.

10. **Demonstrates the Ability of the Device to Cause a Person to Become Charged and Experience a Moderate Shock when Grounded**—Perform as in Fig. 14-11. (Similar to Experiment 3 shown Fig. 14-3). **Caution! Do not attempt on individuals unless you know they are in good physical condition.**

The unit is directed to victim. Ideal condition is for victim to be standing on insulated floor or wearing rubber or other insulative type shoes. Unit should be grounded for maximum effect, however, some grounding effect will occur through person holding it. Use light switch, pipe, etc., for temporary ground.

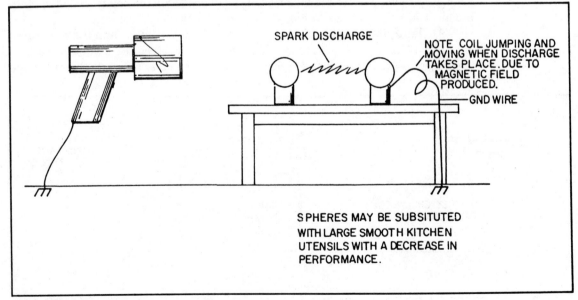

Fig. 14-10. Demonstration of the transmission of energy via mobile ions.

236

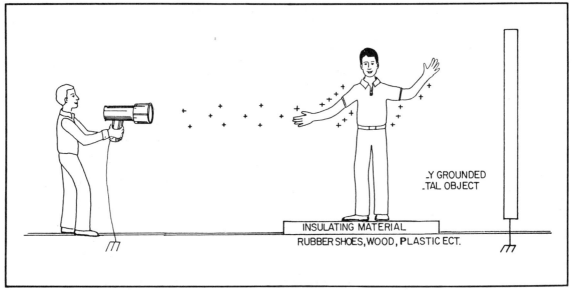

Fig. 14-11. Charging a person with ions (*see cautions!*).

Victim charges up with ions and when he touches grounded or large object, a flow of current occurs causing a shock. **Please note that the severity of this shock depends on many factors and can be severe under proper conditions. Again use Caution!**

Other experiments and uses are: materials and insulation dielectric breakdown testing, X-ray power supplies, capacitance charging, high-voltage work, ignition of gases in tubes and spark gaps, particle acceleration and atom smashers, Kirlian photography, electrostatics and ion generation. Other related material may easily be obtained on the above subjects.

NEGATIVE ION INFORMATION

In the last two decades a medical controversy has evolved pertaining to the beneficial effects of these minute electrical particles. As with any device that appears to affect people in a beneficial sense there are those who sensationalize and exaggerate these claims as a cure for all ailments and ills. Such people manufacture and market these devices under false pretenses, and consequently give the products a bad name. The Food And Drug Administration now steps in on these claims and the product along with its beneficial facets goes down the tubes.

People are affected by negative ions from the property of these particles to increase the rate of activity of cilia (whose property is to keep the tracheas clean from foreign objects) thus enhancing oxygen intake and increasing the flow of mucus. This property neutralizes the effects of cigarette smoking that slows down this activity of the cilia. Hay fever and bronchial asthma victims are greatly relieved by these particles. Burn and surgery patients are relieved of pain and heal faster. Tiredness, lethargy and the general dragged out feeling are replaced by a sense of well being and renewed energy. Negative ions destroy bacteria and purify the air with a country air freshness.

They cheer people up by decreasing the serotonin content of the blood. As can be seen in countless articles and technical writings, negative ions are a benefit to man and his environment. (The reader is encouraged to do further research into this subject at their local library.)

Negative ions occur naturally from static electricity, certain winds, waterfalls, crashing surf, cosmic radiation, radioactivity and ultraviolet radiation. Positive ions are also produced from some of the above phenomena and they usually neutralize each other out as a natural statistical occurrence. However, many manmade objects and devices have a tendency to neutralize the negative ions, thus leaving an abundance of positive ions which create sluggishness and most of the opposite physiological effects of its negative counterpart.

One method of producing negative ions is by obtaining a radioactive source rich in beta radiations (electrons). Alpha and gamma emission from this source produces positive ions that are neutralized electrically. The resulting negative ions are electrostatically directed to the output exit of the device and further dispersed by the action of a fan (this method has recently come under attack by the Bureau Of Radiological Health And Welfare for the use of tritium or other radioactive salts). This approach appears to be the more hazardous of the two according to the Consumer Product Safety People.

A more accepted method is to place a small tuft of stainless steel wool as the *ion emitter* at the output terminal of anegative HV dc power supply. The hair-like property of the stainless steel wool allows ions to be produced at relatively low voltage yet reducing the ozone output. Ions are produced by leakage of the particles charging air molecules in the immediate vicinity of the steel wool emitter. The unit should be operated below 14kV as overvoltage can produce substantial amounts of ozone that can

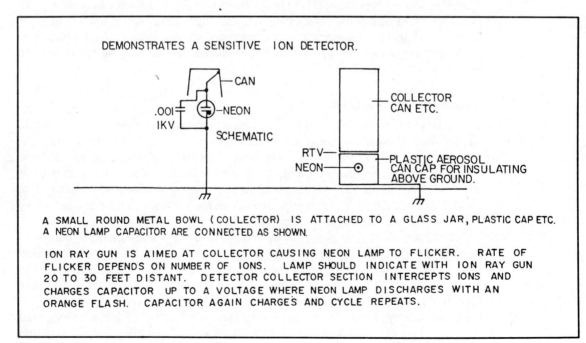

Fig. 14-12. Demonstrates a sensitive ion detector.

mask the beneficial effect of the increased ions obtained. A suitable and effective ion detector capable of indicating relative amounts of ion flux is shown in Fig. 14-12.

Special Note: It is suggestd that the experimenter seriously consider our IOD10, Ion Detector, offered by Information Unlimited, P.O. Box 716, Amherst, NH 03031. This device has many uses in determining static charges in computer rooms, offices etc. It's sensitivity adjustable and responds to both positive and negative. Refer to model IOD10 when ordering.

THEORY OF OPERATION AND CIRCUIT DESCRIPTION

The unit is divided into two sections: The Variable Supply/Oscillator and the Transformer Multiplier (see Fig. 14-13 and Table 14-1). Power is supplied by a grounded three-wire cord (CO1) that plugs into any standard 115-volt ac receptacle. Switch (S1) controls the primary power. The neon lamp (NE1) is illuminated when power is applied through the current-limiting resistor (R1). Variable transformer (VA1) controls the primary voltage to the step-down transformer (T1). T1 supplies low ac voltage to diode bridge (D1,2,3,4) that rectify this voltage and charges up smoothing capacitor (C1) through limit-resistors (R2,3,4).

A variable dc output from 0-16 volts is available by adjustment of VA1. Dc voltage is now applied to the center-tap of the ferrite switching transformer (T2). The collectors of switching transistors (Q1,2) now switch alternate sections of T2 producing a high frequency wave near the resonant frequency of T2. Transistors Q1, 2 are driven by the center top base winding of T2. The driving voltage is fed to these respective transistors to cause a self-sustaining oscillation to occur, hence it must be properly phased to work. Resistor R6 provides a return path for this current, limiting it to a safe value. Resistor R5 supplies the initiating voltage to start oscillation. A high-voltage, high-frequency wave is produced at the output of T2 and fed into a Cockcroft-Walton Voltage Multiplier consisting of the redundant diodes and capacitors, as shown. Output voltage now can be hundreds of kV.

The device is shown built into two sections connected by a short umbilical cord. This approach removes the HV charges produced from the electronics and allows safe remote control. The umbilical wires can be up to 10' to 15', if required. The variable power supply/oscillator section should be built in a shield enclosure and be securely connected to earth ground. Conventional wire and soldering may be used in this section with the only criterion being that of proper fit of the components.

The high frequency transformer is now housed in a nonmetallic enclosure that can be a plastic flower pot. This eliminates leakage and losses due to corona and static-induced currents that would occur near metal. The multiplier section is shown built into an oil-proofed container and requires HV oil for operation. Output is shown via the contact on top of the enclosure. A ground return should be made at T2 ground or J1. This eliminates loop current in the other parts of the system.

ASSEMBLY STEPS

1. Cut a 8-1/2" × 2" wide piece of perfboard as shown in Fig. 14-14.
2. Decide the number of voltage multiplication stages desired. Figure 14-14 shows twelve stages that will produce a voltage output of more than 100 kVdc.

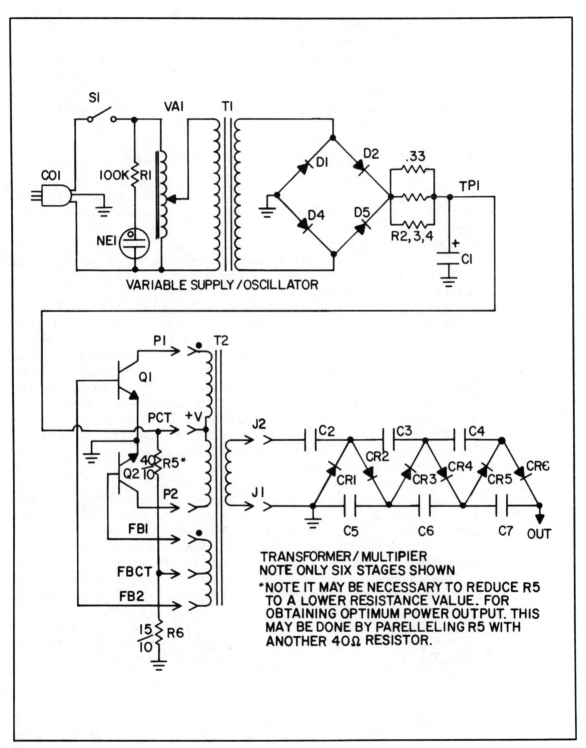

Fig. 14-13. HVGI Schematic basic.

Table 14-1. Parts List (HVG1).

R1	1	100 kΩ 1/4-watt resistor
R2,3,4	3	.33 Ω 5-watt resistor
R5	1	40 Ω 10-watt resistor (See note Fig. 14-13.)
R6	1	15 Ω 10-watt resistor
C1	1	20 to 50,000 ΩF a 25 V electrolytic
*C2 to Cn		.001 ΩF 15 kV
D1-4	4	Diode bridge 25 amps
*CR1 to		
CRn		20 kV avalanche diodes
Q1,2	2	2N3055 NPN power transistor
NE1	1	Neon lamp with leads
S1	1	SPST switch 3 amps
CO1	1	3 wire line cord
VA1	1	175 VA variac
T1	1	16-V/7-amp multitap or any 12-V at 5-amp transformer (Fig. 14-18).
*T2	1	Ferrite flyback transformer (See Fig. 14-19).
MK1,2	2	TO3 mounting kits
PB1	1	Multiplier board 8-1/2" × 2" wide .1 × .1 perfboard
PB2	1	Resistor board 2-3/8" × 2-3/8"
P1,2	2	Banana plugs
J1,2	2	Banana jacks
CA1	1	Rounded PVC cap 3" for 3-1/2" OD
CA2	1	Square PVC cap 3" for 3-1/2" OD
EN1	1	11" by 3" PVC enclosure
CON1	1	Rounded metal contact
EN2	1	7" × 5" × 3" aluminum enclosure
WR1	20'	#20 vinyl wire 20'
WR2	15'	#18 vinyl wire 15'
TYE1	4	8" tyewraps
WR4	1	#16 buss wire
SW1		1/4"-20 × 3/4" nylon screws
SW2		1/4"-20 × 1/2 screws
FEET		Rubber stick on feet
HS1		Heatsink Dual TO3
BU1,2,3,4		1/2" bushing
BU5		Cord clamp bushing
EN3		Plastic pot and base 6"
SCR1		Cover screen fab as needed (Not required.)
CF1		Charge reflector. Use kitchen utensil such as a large metal bowl etc. (Not required for normal operation.)
LAB1		High-voltage label

IOD1—Ion and Field Detector—Small hand-held device indicates the presence of electric fields, ions and other electrical phenomena. A must for those experimenting with negative ions and high voltages.

AS8—Atom Smasher—Full construction plans on building a medium-energy particle accelerator that will split atoms! This is a Lab Quality Device! Plans include full details on a high-voltage multiplier, accelerator assembly and complete high vacuum system.

All parts, assemblies, and completed units are available from Information Unlimited, P.O. Box 716, Amherst, NH 03031 or call (603) 673-4730. Parts marked with an asterisk are individually available.

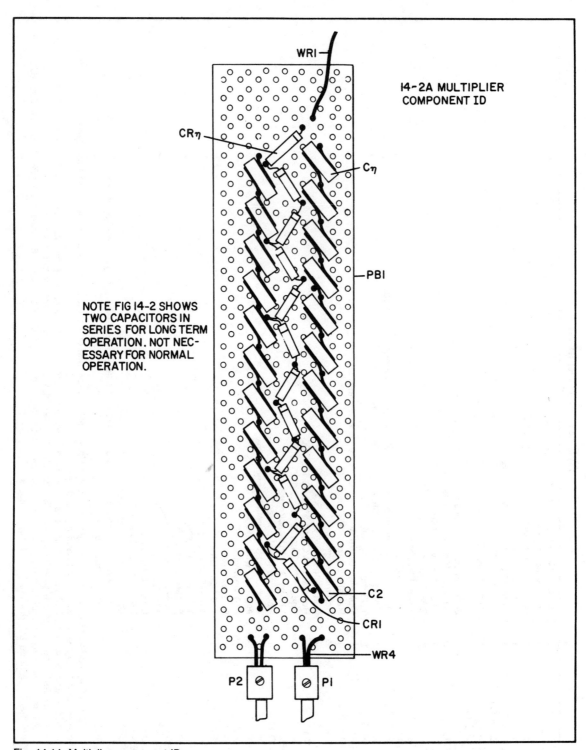

Fig. 14-14. Multiplier component ID.

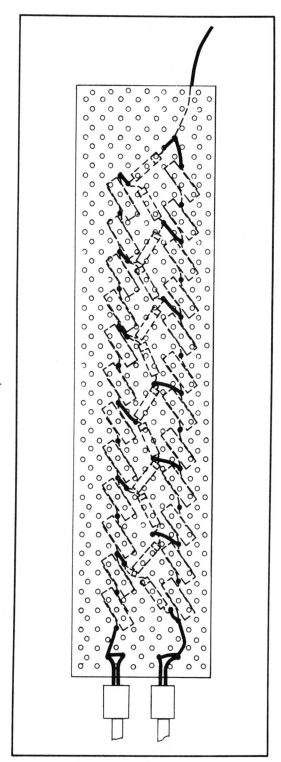

Fig. 14-15. Multiplier wiring aid.

3. Insert components as shown and wire as per Fig. 14-15. Note that solder joints should be round and smooth. Use excess solder to obtain a smooth ball appearance. This is contrary to usual solder methods but is necessary here to prevent damaging corona that will form on any points or irregular surfaces. Use fingers and verify the absence of these points as this is important if device is to be used continuously.

4. Attach banana jacks as shown using wire leads passed through perfboard to secure. The method we use is to double wire these two holes in perfboard using #18 brass wire for plug attaching to attach output wire (WR1).

5. Assemble resistor board as shown in Fig. 14-16. Follow approximate layout.

6. Assemble heat sink and power transistor as shown in Fig. 14-17. Note inset drawing on proper assembly of TO3 mounting kits.

7. Wire transformer T1 as shown in Fig. 14-18. This wiring is intended for the transformer specified in Table 14-1.

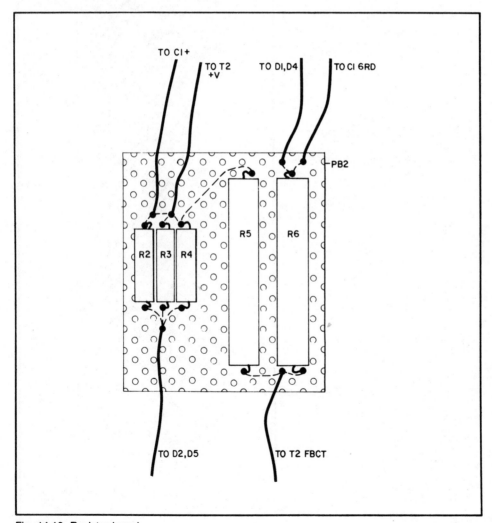

Fig. 14-16. Resistor board.

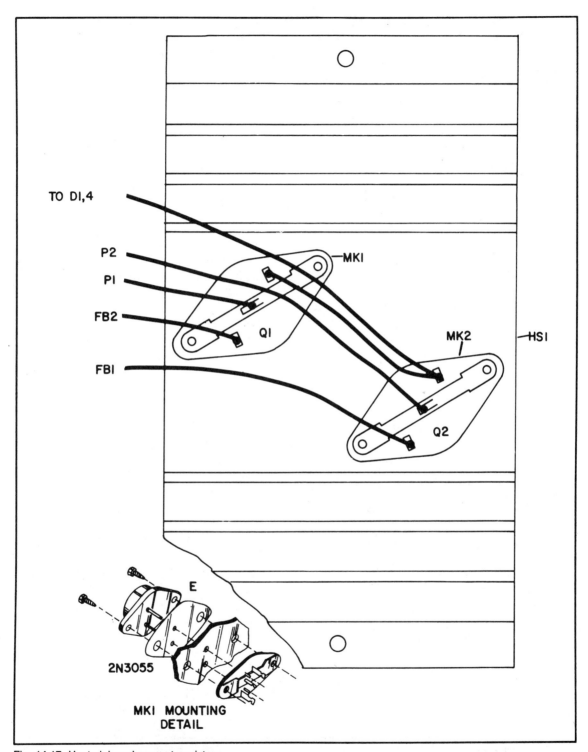

Fig. 14-17. Heat sink and power transistors.

8. Assemble flyback transformer T2 as shown in Fig. 14-19. Use a long length of wire that will allow a 4-foot separation between the multiplier and power supply.

9. Trial position all components and verify fit per Figs. 14-20 through 14-22. Mark for drilling holes. Fabricate as shown and remove all metal burrs and sharp edges. Watch for correct polarity of C1 and diode bridge D1-4. Note that variac VA1 should have insulated pieces of plastic between it and enclosure top and bottom to prevent contact.

10. Wire in components using the schematic in Fig. 14-13 and Fig. 14-20. Use #18 wire for low-voltage leads and #20 for the primary leads from VA1 to T1. Note umbilical wires for T2 with connection identification P1, P2, PCT, FB1, FB2, FBCT shown in Figs. 14-13 and 14-17. Be very careful not to pinch the wires shown in Fig. 14-4 between the heat-sink fins and enclosure.

11. Verify wiring and soldering to this point. Turn VA1 full ccw and S1 to "off." Connect a voltmeter to TP1 and set to 50-volt range. Turn S1 "on" and note pilot lamp NE1 glowing. Turn A1 slowly cw and note voltmeter reading increasing from 0 to 15 volts. Touch output wire of T2 with an insulated metal screwdriver. Note a healthy arc being drawn nearly 1″ in length. Turn VA1 ccw and note arc diminishing. Note if no output is obtained at T2 it may be necessary to reverse TB1 and TB2 at the transistor sockets (Fig. 14-17).

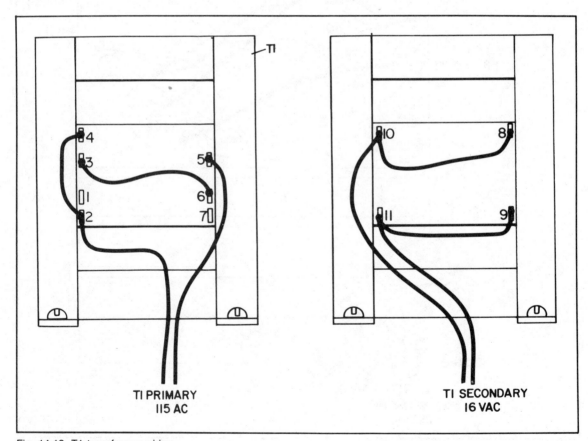

Fig. 14-18. T1 transformer wiring.

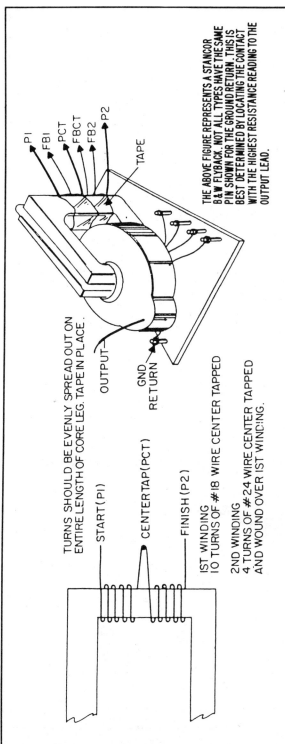

THE ABOVE FIGURE REPRESENTS A STANCOR B & W FLYBACK. NOT ALL TYPES HAVE THE SAME PIN SHOWN FOR THE GROUND RETURN. THIS IS BEST DETERMINED BY LOCATING THE CONTACT WITH THE HIGHEST RESISTANCE READING TO THE OUTPUT LEAD.

TURNS SHOULD BE EVENLY SPREAD OUT ON ENTIRE LENGTH OF CORE LEG. TAPE IN PLACE.

START (PI)

CENTER TAP (PCT)

FINISH (P2)

IST WINDING
10 TURNS OF #18 WIRE CENTER TAPPED

2ND WINDING
4 TURNS OF #24 WIRE CENTER TAPPED
AND WOUND OVER IST WINDING.

WINDING I

TWO NEW WINDINGS ("PRIMARY" AND "FEEDBACK" WINDINGS) ARE ADDED TO THE FLYBACK TRANSFORMER (T2) THAT CONNECTS TO THE DRIVER TRANSISTORS. THESE WINDINGS ARE HAND WOUND ON THE BOTTOM LEG OF THE FERRITE CORE WHERE THE ORIGINAL TWO TURN FILAMENT WINDING WAS LOCATED. REMOVE AND DISCARD THE ORIGINAL FILAMENT WINDING, IN ITS PLACE, WIND FIRST A TEN-TURN, CENTER-TAPPED WINDING (DESIGNATED PI-P2) USING PROPER LENGTH OF No. 18 WIRE OR LARGER INSULATED HOOKUP WIRE. THIS IS EASILY ACCOMPLISHED BY WINDING FIVE TURNS AT ONE END OF THE CORE AND THEN TWISTING A LOOP IN THE FREE END BEFORE ADDING THE SECOND FIVE TURNS. THE COMPLETE TEN TURN WINDING SHOULD THEN BE HELD IN PLACE WITH A TURN OR TWO OF ELECTRICAL TAPE WITH THE TWO ENDS (PI & P2) AND THE CENTER TAP (PCT) LOOP ALL PROTRUDING. CONNECTION CAN BE MADE TO THE CENTER TAP LOOP WHEN THE INSULATION HAS BEEN CAREFULLY REMOVED. IF IT BECOMES NECESSARY TO CUT THE CENTER LOOP, BE SURE THAT THE TWO ENDS ARE SCRAPED AND JOINED TO FORM A MECHANICAL AS WELL AS AN ELECTRICAL CENTER TAP CONNECTION TO THE WINDING.

WINDING II

THE SECOND WINDING (FEEDBACK) SHOULD BE WOUND DIRECTLY ON TOP OF THE FIRST, BUT IT SHOULD ONLY HAVE A TOTAL OF FOUR TURNS—TWO EACH SIDE OF THE CENTER TAP. WIND TWO TURNS OF #22 HOOK-UP WIRE, PULL AND TWIST A CENTER TAP LOOP (FBCT) AND WIND THE OTHER TWO TURNS. TAPE THIS WINDING IN PLACE ON TOP OF THE FIRST. DO NOT LET THE CENTER TAP LOOPS OF THE TWO WINDINGS TOUCH EACH OTHER

Fig. 14-19. T2 winding instructions.

247

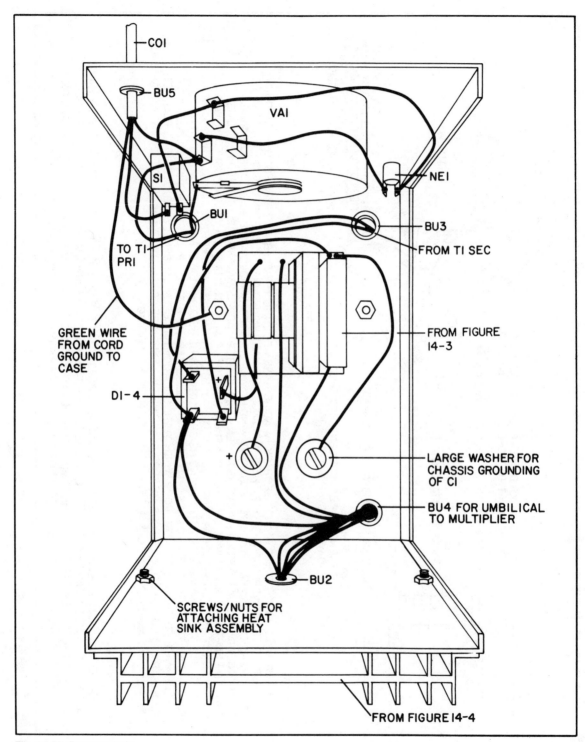

Fig. 14-20. Power supply.

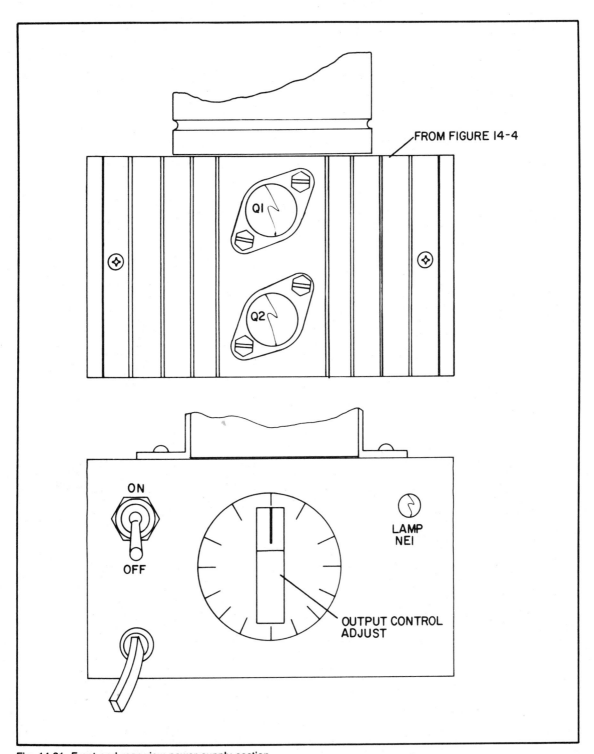

Fig. 14-21. Front and rear view power supply section.

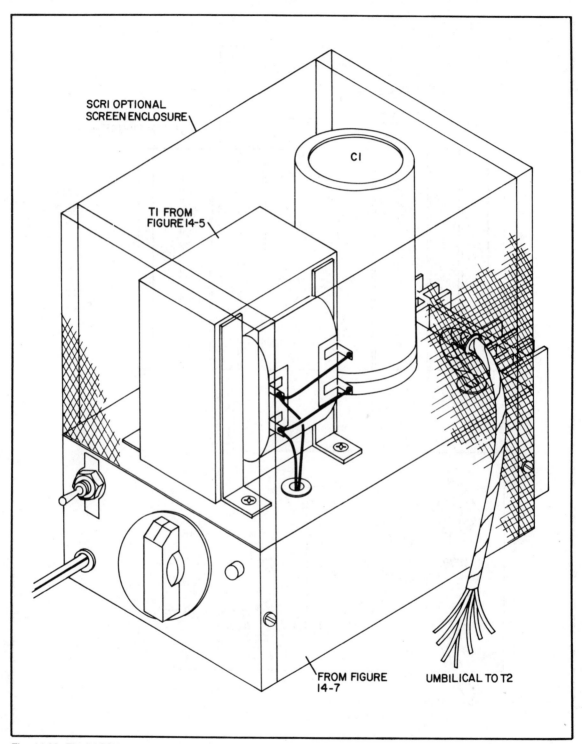

SCRI OPTIONAL
SCREEN ENCLOSURE

CI

TI FROM
FIGURE 14-5

FROM FIGURE
14-7

UMBILICAL TO T2

Fig. 14-22. Final ASSY power supply.

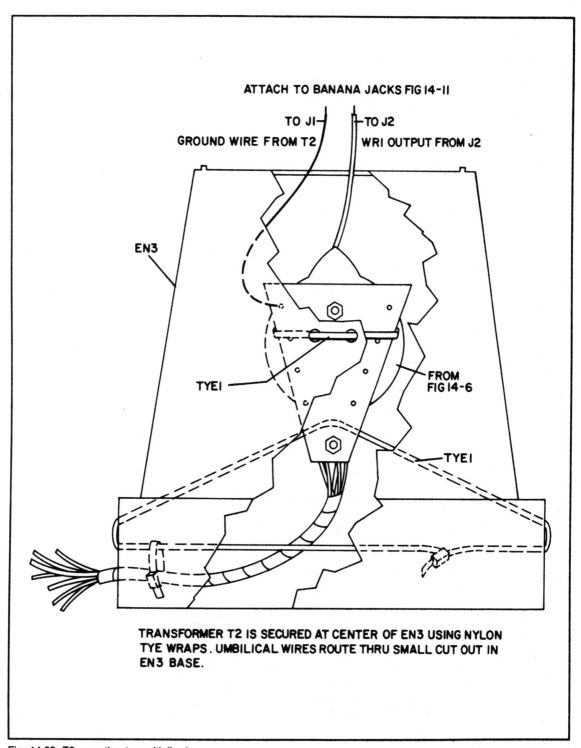

ATTACH TO BANANA JACKS FIG 14-11

TO J1
GROUND WIRE FROM T2

TO J2
WRI OUTPUT FROM J2

EN3

TYE1

FROM FIG 14-6

TYE1

TRANSFORMER T2 IS SECURED AT CENTER OF EN3 USING NYLON TYE WRAPS. UMBILICAL WIRES ROUTE THRU SMALL CUT OUT IN EN3 BASE.

Fig. 14-23. T2 mounting to multiplier base.

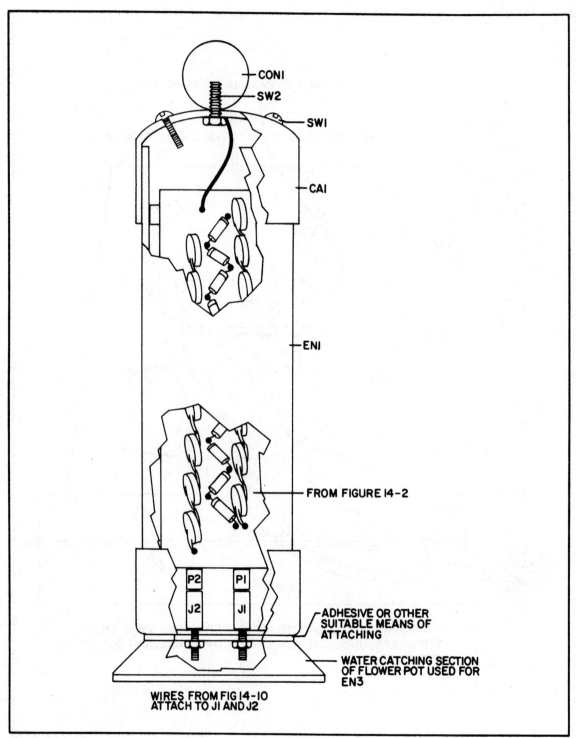

Fig. 14-24. X-ray multiplier.

12. Fabricate EN3 (Fig. 14-23) from a plastic 6″ flower pot. These are readily available from hardware or flower shops. This pot comes with a water catcher section as shown in Fig. 14-24. T2 is suspended at the center of EN3 using tyewraps secured to the sides via small holes. Use ingenuity in performing this step. Note the output wire of T2 routing up the top of EN3. These wires will eventually be attached to J1 and J2 of the multiplier (Fig. 14-24).

13. Fabricate EN1 from an 11″ × 3-1/2″ OD piece of PVC tube.

14. Fabricate square cap CA2 by placing two holes for jacks J1 and J2. Note that these must be identically spaced for mating with plugs P1 and P2 from Fig. 14-14. Seal around jacks with a two part epoxy to keep oil from leaking out. Attach to EN1 using necessary PVC cleaners and sealers for an oil-proof seal. It should now be easy to plug in multiplier board (Fig. 14-14) into mating jacks. Note that these connections are polarized—make note of ground and T2 output.

15. Fabricate CA1 from a preferably rounded cap (square may be used but doesn't look as nice) by drilling a 1/4″ hole in the center for attaching brass ball (CON1). Drill and tap two 1/4″-20 holes as shown in Fig. 14-24 for oil fill and vent. It is not necessary

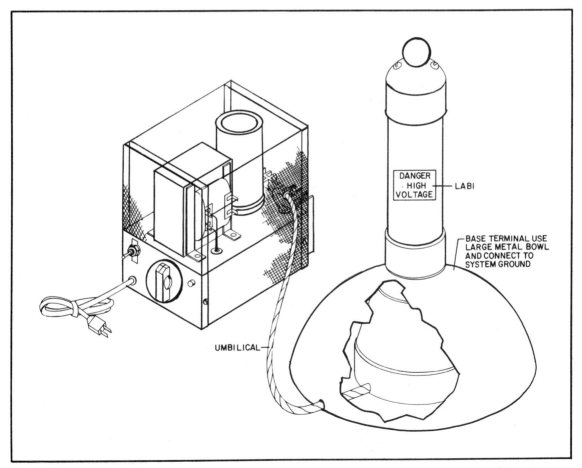

Fig. 14-25. Finished system.

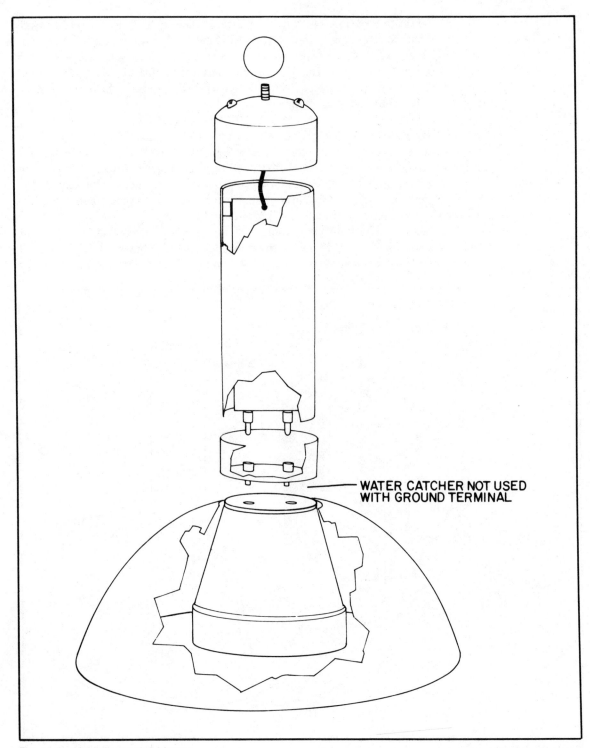

WATER CATCHER NOT USED
WITH GROUND TERMINAL

Fig. 14-26. Multiplier section blowup.

to drill two holes if multiplier is filled before installing to cap CA1. Note that it is not advisable to seal the top cap to the tube permanently as this makes the whole section useless should something inside break down. Note output wire of multiplier must be connected to output terminal.

16. Attach water catcher section of flower pot to bottom of CA2 (Fig. 14-24) using epoxy or other suitable glue. Note holes necessary for clearance of J1 and J2. The wires from T2 can now be attached. They must be as short as possible and go to the correct jack. The multiplier section can now be placed on EN3 enclosure without any attachment between the two. It is advisable however to use some type of securing such as double sided tape placed between the two. This helps prevent the unit from toppling over and possibly spilling oil all over the place.

17. Place finalized assembly into the position shown in Fig. 14-25. (Also see Fig. 14-26.) Attach HV label LAB1 as shown. Attach a sturdy ground wire to the power supply section via one of the screws. This wire should be attached to a well-insulated metal object.

18. Fill container almost to the top with oil and put on the "top cap" CA1. Attach wire to output using ingenuity.

19. Slowly apply power to midscale on VA1 and place grounded contact near output. Note a healthy discharge about 2 to 3" in length.

20. Your system is now ready to operate. It is not a good idea to run the unit at full power for long periods of time or dangerous amounts of ozone may be generated. Use in a well-ventilated area. An outer PVC sleeve may be extended to the present one to provide a shielded and recessed output terminal.

Index

Index

Edited by Roland S. Phelps

Other Bestsellers of Related Interest

THE LASER COOKBOOK:
88 Practical Projects—Gordon McComb

The laser is one of the most important inventions to come along this half of the 20th Century. This book provides 88 laser-based projects that are geared toward the garage-shop tinkerer on a limited budget. The projects vary from experimenting with laser optics and constructing a laser optical bench to using lasers for light shows, gunnery practice, even beginning and advanced holography. 400 pages, 356 illustrations. Book No. 3090, $19.95 paperback, $25.95 hardcover

50 CMOS IC PROJECTS—Delton T. Horn

Delton T. Horn presents a general introduction to CMOS ICs and technology . . . provides full schematics including working diagrams and parts lists . . . offers construction hints as well as suggestions for project variations and combinations. This book discusses: the basics of digital electronics, safe handling of CMOS devices, breadboarding, tips on experimenting with circuits, and more. You'll find signal generator and music-making projects, time-keeping circuits, game circuits, and a host of other miscellaneous circuits. 224 pages, 226 illustrations. Book No. 2995, $25.95 hardcover only

THE ROBOT BUILDER'S BONANZA:
99 Inexpensive Robotics Projects
—Gordon McComb

Where others might only see useless surplus parts, you can imagine a new "life form." With the help of this book, you can truly express your creativity. It offers you a complete, unique collection of tested and proven product modules that you can mix and match to create an almost endless variety of highly intelligent and workable robot creatures. 336 pages, 283 illustrations. Book No. 2800, $17.95 paperback only

SUPERCONDUCTIVITY: The Threshold of
a New Technology—Jonathan L. Mayo

Superconductivity is generating an excitement not seen in the scientific world for decades! Experts are predicting advances in state-of-the-art technology that will make most existing electrical and electronic technologies obsolete! This book is one of the most complete and thorough introductions to a multifaceted phenomenon that covers the full spectrum of superconductivity and superconductive technology. 160 pages, 58 illustrations. Book No. 3022, $12.95 paperback only

FIBEROPTICS AND LASER HANDBOOK
—2nd Edition—Edward L. Safford, Jr., and
John A. McCann

Explore the dramatic impact that lasers and fiberoptics have on our daily lives—PLUS, exciting ideas for your own experiments! Now, with the help of experts Safford and McCann, you'll discover the most current concepts, practices, and applications of fiberoptics, lasers, and electromagnetic radiation technology. Included are terms and definitions, discussions of the types and operations of current systems, and amazingly simple experiments you can conduct! 240 pages, 108 illustrations. Book No. 2981, $18.95 paperback only

BUILD YOUR OWN LASER, PHASER, ION RAY
GUN & OTHER WORKING SPACE-AGE
PROJECTS—Robert E. Iannini

Here's the highly-skilled do-it-yourself guidance that makes it possible for you to build such interesting and useful projects as a burning laser, a high power ruby YAG, a high-frequency translator, a light beam communications system, a snooper phone listening device, and more—24 exciting projects in all! 402 pages, 302 illustrations. Book No. 1604, $17.95 paperback only

GORDON McCOMB'S GADGETEER'S GOLDMINE!: 55 Space-Age Projects
—Gordon McComb

If you're into (or are ready to try) high-tech experiments with lasers, fiber optics, power supplies, high-voltage devices, and robotics, this is the book for you. It features 55 practical projects ranging in difficulty from a simple Jacob's Ladder to a complete laser light show. All designs have been thoroughly tested, and suggested alternative approaches, parts lists, sources, and components are provided. 432 pages, 274 illustrations. Book No. 3360, $18.95 paperback, $29.95 hardcover

GORDON McCOMB'S TIPS & TECHNIQUES FOR THE ELECTRONICS HOBBYIST
—Gordon McComb

This concise handbook covers everything from setting up a shop to making essential equipment. You'll find general information on electronics practice, important formulas, tips on how to identify components, and more. Use it for ideas, as a text on techniques and procedures, and as a databook on formulas, functions, and components. 288 pages, 307 illustrations. Book No. 3485, $17.95 paperback, $27.95 hardcover